ENCYCLOPAEDIA OF MINERALOGY

ENCYCLOPAEDIA OF MINERALOGY

Volume 1

Bharat Singh

ANMOL PUBLICATIONS PVT. LTD.
NEW DELHI-110 002 (INDIA)

ANMOL PUBLICATIONS PVT. LTD.
Regd. Office: 4360/4, Ansari Road, Daryaganj,
New Delhi-110002 (India)
Tel.: 23278000, 23261597, 23286875, 23255577
Fax: 91-11-23280289
Email: anmolpub@gmail.com
Visit us at: www.anmolpublications.com

Branch Office: No. 1015, Ist Main Road, BSK IIIrd Stage
IIIrd Phase, IIIrd Block, Bengaluru-560 085 (India)
Tel.: 080-41723429 • Fax: 080-26723604
Email: anmolpublicationsbangalore@gmail.com

Encyclopaedia of Mineralogy

ISBN 978-81-261-3475-5 (Set)

PRINTED IN INDIA

Printed at AnVi Composers, New Delhi.

Contents

Preface

'Mineralogy' is a branch of science, under which we study about the minerals, which are obtained from the earth. All the minerals are found in the earth. We dig out the earth for getting minerals in the form of ores, which are impure. By the process of Metallurgy, we get pure elements. A few metals such as Au (gold), Ag (silver) and Pt (platinum) are found in pure form. A mineral is defined to be a naturally occurring, homogeneous crystalline solid. First, a mineral should be a solid; this requirement excludes all liquids and gases. A crystalline material is not merely a solid, but also possesses an orderly and periodic molecular structure. In recent years, however, a multitude of synthetic materials, which would otherwise satisfy the definition of a mineral, have been produced in the laboratory. These substances are rapidly gaining in importance in both industry and experiment, and it now seems logical to extend the definition of 'mineral'. Minerals are inorganic solids, which have a regular three-dimensional ordered structure of atoms.

Minerals are classified on the basis of their chemical composition and their isomorphism or similarity of properties, such as structural properties, physical properties and chemical properties.

In structural properties, we study about the structure of the crystal. Crystallographic property is one of the most characteristic properties of the crystals, and in this property we study about the shape of the crystals. Other structural property is 'optical properly', in which we study about the crystal structure, the kind of atoms present in the structure and their electronic arrangement. In physical properties, we study about the properties such as colour, taste, melting and boiling points, density, etc. In chemical properties, we study about the reactions of the elements/minerals with others and products after reactions and their relation with the reactants.

This eminent work, *Encyclopaedia of Mineralogy* has specific and realistic features about minerals in an interesting manner, which will be able to make a creative impression in the study of mineralogy. Expectedly, this work would be welcomed and liked by all.

— **Editor**

1

Introduction

Mineralogy is an earth science focused around the chemistry, crystal structure, and physical (including optical) properties of minerals. Specific studies within mineralogy include the processes of mineral origin and creation (along with destruction), classification of minerals, their geographical distribution, as well as their utilisation.

A mineral is defined to be a naturally occurring, homogeneous crystalline solid. A mineral must first of all be a solid; this requirement excludes all liquids and gases. A crystalline material is not merely a solid but also possesses an orderly and periodic molecular structure. This molecular structure is composed of repeated units, each of which contains a specified set of atoms or ions placed according to a specified geometric arrangement.

The periodic repetition of such unit cells results in discrete translational invariance within the crystal lattice. An amorphous solid such as amber, coal, or glass does not possess this orderly, periodic, translationally invariant crystalline lattice structure and thus may not be considered a mineral.

Minerals are expected to be homogeneous: the chemical composition of a mineral may not differ significantly from sample to sample within a species or from location to location within a sample. However, small variations in chemical composition are permissible. In certain species the chemical composition may vary over a well-defined range. Two samples, each of which possesses a chemical composition which falls within the required range, are then considered to be minerals of the same species despite any difference in their compositions. Trace amounts of impurities, or elements which are not an integral part of the crystal structure, may also be present within the crystal lattice. The atoms of an impurity may either occupy a previously uninhabited position within a few individual unit cells or they may replace a few atoms which are regularly part of the lattice. Such impurities may not be present in large quantities and they may not greatly affect any physical properties of the mineral species except colour, hardness, and conductivity.

The requirement that a mineral be naturally occurring was conceived before the production of synthetic materials was prevalent. In recent years, however, a multitude of synthetic materials which would otherwise satisfy the definition of a mineral have been produced in the laboratory. These substances are rapidly gaining in importance in both industry and experiment, and it now seems logical to extend the definition of 'mineral' in order to include such synthetic materials.

The natural sciences geology and geophysics in first instance study the formation of the earth, but nowadays also extraterrestrial bodies such as planets in and outside our solar system. In general geology focuses on large scale processes such the formation of mountain ranges and movements in the earth crust. In contrast, mineralogy focuses on the elemental parts that form the rocks and that determine their general appearance, the minerals. Nowadays, mineralogy is no longer confined to terrestrial material but also to materials of extraterrestrial origin. In the last decades the nature of mineralogy has drastically changed and can no longer easily be defined. In general one can say with certainty that, in addition to being a branch of geology, it is a branch of inorganic chemistry, but the discipline focuses on the origin, description, and classification of minerals. Nowadays, four main categories can be described;

(1) Crystal chemistry (composition, atomic arrangements)

(2) Paragenic mineralogy (study of mineral associations and occurrences)

(3) Descriptive mineralogy (study of physical properties and means for identification)

(4) Taxonomic mineralogy (classification, systematisation and nomenclature)

The word mineral comes from the Latin word *minare,* which means mining. It expresses the strong historical relationship between mining and the history of studying the ore, the minerals. Nowadays research in mineralogy is much more oriented to the formation of minerals, under which physical and chemical conditions, their behaviour under high pressure and temperature, their crystallography, their chemical composition and their possible uses.

Minerals are defined as inorganic solids consisting of a regular three-dimensional ordered structure of atoms. This ordering of the atoms determines not only the crystalline nature of minerals, but it also gives us the opportunity to give a mineral a characteristic chemical composition. When minerals can grow without any hindrance they will be terminated by crystal faces, which are ordered according to a regular system resulting in a characteristic relationship typical for each mineral. This is studied in the crystallography.

Classification of Minerals

Minerals are classified firstly based on their chemical composition, and then so far as possible by isomorphism or similarity of crystalline form. In general the cations are of less importance than the anions or anionic groups. Normally the following ten major groups are identified

		Characteristic Anion or Anionic Group
I	native elements	
II	sulphides	S
III	oxides and hydroxides	O, OH
IV	halogen minerals	Cl, F
V	carbonates	CO_3
VI	nitrates, borates	NO_3, BO_3, B_4O_7, B_4O_{11}, B_5O_9
VII	sulphates, chromates	SO_4, CrO_4, Cr_2O_7
VIII	tungstates, molybdates	WO_4. MoO_4
IX	phosphates, arsenates, vanadates	PO_4, AsO_4, VO_4
X	silicates	Si_2O_4

Mineral Names

Minerals usually have both mineral name and a chemical name. Thus for example silicon dioxide is generally known as quartz, or carbon is, depending on the crystal structure, either known as graphite or as diamond.

Some old mineral names are of unknown origin, while many come from Latin words, e.g. orpiment (auri pigmentum) or from Greek words, e.g. chalcosite (chalcos = copper or bras) or saponite (sapoun = soap).

Most modern minerals names end with "ite". Some have a chemical annotation, such as molybdenite (Molybdenum sulphide MoS_2) or zincite (zinc oxide ZnO). Minerals like tetrahedrite and hemimorphite have names with a crystallographic origin. Specific physical properties also form the basis for mineral names, e.g. magnetite (magnetic), graphite (to write), rhodochrosite (rose colour). Geographic origin forms sometimes the basis for mineral names, such as in labradorite and vesuvianite. Last but not least, a large number of minerals have been named after individuals, e.g. scheelite, smithsonite, goethite, etc.

Properties of Minerals

Crystallographic Properties: One of the most characteristic properties of crystals is the shape of their crystals (also known as morphology). When crystals can grow freely they will clearly show their specific crystal faces (idiomorphic crystals). However, in many instances crystals will be deformed, e.g. due to lack of space or interference with other growing crystals. Also, not always the same faces will be recognised on a crystal. For example in a cubic system one can observe the cubic faces, but also octaeder or rhombododecaeder faces or combinations of faces in a single crystal. This can result in large differences in morphology. Analysis of the symmetry relations of the faces enables a mineralogist to determine its crystal class.

Optical Properties: Optical properties of minerals are related to the crystal structure, the kind of atoms present in the structure and their electronic arrangement. The optical properties are usually examined under a microscope by examining small fragments in transmitted plane-polarised light for non-opaque minerals and in reflected plane-polarised light for opaque minerals. Amorphous and cubic substances are isotropic, they have only one index of refraction for give monochromatic light. Hexagonal and tetragonal minerals are uniaxial and are also double refracting. All other minerals are both biaxial and double refracting. The optical properties of most minerals are well known and are extensively tabulated in books and other publications. In certain instances the optical properties are related to the chemical composition of a mineral series, e.g. the plagioclase series. The optical properties are also related to the colour of the mineral.

Physical Properties: In general the physical properties of a mineral are those properties that can be used for the identification by sight, touch, etc. These properties are directly related to the crystal structure, the chemical composition and the kinds of chemical bonds present in the crystal structure. The two best defined properties are hardness, cleavage and density. Colour and lustre are more difficult to access.

The hardness of a mineral is basically related to the chemical bonds present in the crystal structure. So, minerals with stronger chemical bonds are harder than those with weaker bonds. Absolute hardness can be measured by a microhardness tester, which places an indentation on a surface of the surface of a crystal beyond a certain pressure value. Much easier to use, especially in the field, for mineralogist is a relative scale based on 10 minerals with increasing hardness. This scale is known as the Mohs scale.

1	talc	6	potash feldspar
2	gypsum	7	quartz
3	calcite	8	topaz
4	fluoride	9	corundum
5	apatite	10	diamond

Thus a mineral with a hardness of 6.5 will scratch potash feldspar but will not scratch quartz.

Cleavage is the preferential breakage along certain crystallographic planes. Cleavage is easiest along those planes with the weakest chemical bonds in the crystal structure. If all bonds are similar in all three directions (e.g. in framework silicates) then the plane with the least bonds per unit area will be the preferred cleavage plane. Cleavage is generally classified based on the perfection of the cleavage: perfect, good, distinct, poor or no cleavage. Fracture is the way a mineral breaks other than cleavage. Typical terms used for fractures are splintery, hackly, conchoidal, rough and smooth.

Density plays a significant role in the crystallochemical calculations as it is directly related to the structure and the chemical formula. Density is normally expressed in units of grams per cubic centimetre. Closely related to density is the specific gravity. The specific gravity is the ratio of the weight of a mineral to the weight of an equal volume of water (theoretically at 4°C).

Colour is more difficult to access. Quite a number of

minerals may occur in one, two or many different colours, but usually pure end-members (which are hardly found in nature, but only in synthetic minerals) have a particular colour. The colour of a mineral is caused by the adsorption of certain wavelengths of light. The colour corresponds to the least absorbed wavelengths.

One can make a distinction between idiochromatic minerals, which have their own specific colour, and allochromatic minerals, where the colour is not determined by the mineral itself. Colour is often caused by impurities or crystal imperfections. Careful description of the colour of a mineral can be obtained by comparison to standard colour charts. Streak is the colour of finely powdered mineral particles and is usually obtained by scratching the mineral over an unglazed porcelain surface.

Lustre is very subjective and describes basically the appearance of the mineral surface in reflected light and is related to the index of refraction in a direct way. Lustre increases with increasing index of refraction. Two important types of lustre are metallic and non-metallic. Many elements such as silver and sulphides are metallic in appearance. Non-metallic lustre can be divided in vitreous (e.g. quartz), resinous (e.g. sphalerite) or adamantine (e.g. diamond). The lustre of a mineral is also influenced by its physical appearance or morphology. Fibrous minerals may have a silky lustre, while platy minerals often display a pearly lustre.

Fluorescence is observed when a mineral that absorbs ultraviolet light, emits visible light as a result. The word fluorescence comes from the mineral fluoride, which shows this property very beautifully. Although interesting and sometimes spectacular, this property is only useful in very special occasions for identification purposes. Fluorescence is not a constant property and can vary from sample to sample and from locality to locality. Other properties that sometimes can be used for identification are magnet, radioactive or electrical properties.

History: Early speculation, study, and theory of mineralogy was written of in ancient Babylonia, the ancient Greco-Roman world, ancient and medieval China, and noted in the prana of Sanskrit texts from ancient India. However, the first systematic scientific studies of minerals and rocks was a development of post-Renaissance Europe. The credible study of mineralogy was founded on the principles of crystallography and microscopic study of rock sections with the invention of the microscope in the 17th century.

Europe and the Middle East

The metaphysical emanation and exhalation (*anathumiaseis*) theory of the Greek philosopher Aristotle (384 BC – March 7, 322 BC) included early speculation on earth sciences including mineralogy. According to his theory, while metals were supposed to be congealed by means of moist exhalation, dry gaseous exhalation (*pneumatodestera*) was the efficient material cause of minerals found in the earth's soil. He postulated these ideas by using the examples of moisture on the surface of the earth (a moist vapour 'potentially like water'), while the other was from the earth itself, pertaining to the attributes of hot, dry, smoky, and highly combustible ('potentially like fire'). Aristotle's metaphysical theory from times of antiquity had wide-ranging influence on similar theory found in later medieval Europe, as the historian Berthelot notes:

> *The theory of exhalations was the point of departure for later ideas on the generation of metals in the earth, which we meet with Proclus, and which reigned throughout the middle ages.*

With philosophers such as Proclus, the theory of Neo-platonism also spread to the Islamic world during the middle ages, providing a basis for metaphysical ideas on mineralogy in the medieval Middle East as well. The medieval Islamic scientists expanded upon this as well, including the Persian scientist Ibn Sina (980-1037 AD), who rejected alchemy and the earlier notion of Greek metaphysics that metallic and other elements could be transformed into one another. However, what was largely accurate of the ancient Greek and medieval metaphysical ideas on mineralogy was the slow chemical change in composition of the earth's crust. There was also the Islamic scientist Jabir ibn Hayyan (721-815 AD), who was the first to bring experimental method into alchemy. Aided by Greek Pythagorean mathematics, he discovered the syntheses for hydrochloric acid, nitric acid, distillation, and crystallisation (the latter two being essential for the understanding of modern mineralogy).

Ancient Greek terminology of minerals has also stuck through the ages with widespread usage in modern times. For example, the Greek word asbestos (meaning 'inextinguishable', or 'unquenchable'), for the unusual mineral known today containing fibrous structure. The ancient historians Strabo (63 BC-19 AD) and Pliny the Elder (23-79 AD) both wrote of asbestos, its qualities, and its origins, with the Hellenistic belief that it was of a type of vegetable. Pliny the Elder listed it as a mineral common in India, while the historian Yu Huan (239-265 AD) of China listed this 'fireproof cloth' as a product of ancient Rome or Arabia (Chinese: Daqin). Although documentation of these minerals in ancient times does not fit the manner of modern scientific classification, there was nonetheless extensive written work on early mineralogy. For example, Pliny devoted 5 entire volumes of his work *Naturalis Historia* (77 AD) to the classification of "earths, metals, stones, and gems".

Georgius Agricola, 'Father of Mineralogy'

In the early 16th century AD, the writings of the German scientist George Bauer, pen-name Georgius Agricola (1494-1555 AD), in his *Bermannus, sive de re metallica dialogus* (1530) is considered to be the official establishment of mineralogy in the modern sense of its study. He wrote the treatise while working as a town physician and making observations in Joachimsthal, which was then a centre for mining and metallurgic smelting industries. In 1544, he published his written work *De ortu et causis subterraneorum,* which is considered to be the foundational work of modern physical geology. In it (much like Ibn Sina) he heavily criticised the theories laid out by the ancient Greeks such as Aristotle. His work on mineralogy and metallurgy continued with the publication of *De veteribus et novis metallis* in 1546, and culminated in his best known works, the *De re metallica* of 1556. For the next two centuries this written work remained the authoritative text on mining in Europe.

Agricola had many various theories on mineralogy based on empirical observation, including understanding of the concept of ore channels that were formed by the circulation of ground waters ('succi') in fissures subsequent to the deposition of the surrounding rocks. As will be noted below, the medieval Chinese previously had conceptions of this as well.

For his works, Agricola is posthumously known as the "Father of Mineralogy".

China and the Far East

Unlike Ibn Sina, the ancient Chinese used attributes of alchemy to describe early conceptions of mineralogy. However, medieval Chinese writers provided additional and more complex theories of mineralogy.

Chinese ideas of metaphysical mineralogy span back to at least the ancient Han Dynasty (202 BC-220 AD). From the 2nd century BC text of the *Huai Nan Zi,* the Chinese used ideological Taoist terms to describe meteorology, precipitation, different types of minerals, metallurgy, and alchemy. Although the understanding of these concepts in Han times was Taoist in nature, the theories proposed were similar to the Aristotelian theory of mineralogical exhalations. By 122 BC, the Chinese had thus formulated the theory for metamorphosis of minerals, although it is noted by historians such as Dubs that the tradition of alchemical-mineralogical Chinese doctrine stems back to the School of Naturalists headed by the philosopher Zou Yan

(305 BC-240 BC). Within the broad categories of rocks and stones (shi) and metals and alloys (jin), by Han times the Chinese had hundreds (if not thousands) of listed types of stones and minerals, along with theories for how they were formed.

In the 5th century AD, Prince Qian Ping Wang of the Liu Song Dynasty wrote in the encyclopedia *Tai-ping Yu Lan* (circa 444 AD, from the lost book *Dian Shu,* or *Management of all Techniques*) (Wade-Giles spelling):

> The most precious things in the world are stored in the innermost regions of all. For example, there is orpiment (tzhu huang). After a thousand years it changes into realgar (hsiung huang). After another thousand years the realgar becomes transformed into yellow gold.

In ancient and medieval China, mineralogy became firmly tied to empirical observations in pharmaceutics and medicine. For example, the famous horologist and mechanical engineer Su Song (1020-1101 AD) of the Song Dynasty (960-1279 AD) wrote of mineralogy and pharmacology in his *Ben Cao Tu Jing* of 1070. In it he created a systematic approach to listing various different minerals and their use in medicinal concoctions, such as all the variously known forms of mica that could be used to cure various ills through digestion. Su Song also wrote of the subconchoidal fracture of native cinnabar, signs of ore beds, and provided description on crystal form. Similar to the ore channels formed by circulation of ground water mentioned above with the German scientist Agricola, Su Song made similar statements concerning copper carbonate, as did the earlier *Ri Hua Ben Cao* of 970 AD with copper sulphate.

The Yuan Dynasty scientist Zhang Si-xiao (died 1332 AD) provided a groundbreaking treatise on the conception of ore beds from the circulation of ground waters and rock fissures, two centuries before Georgius Agricola would come to similar conclusions. In his *Suo-Nan Wen Ji,* he applies this theory in describing the deposition of minerals by evaporation of (or precipitation from) ground waters in ore channels.

In addition to alchemical theory posed above, later Chinese writers such as the Ming Dynasty physician Li Shizhen (1518-1593 AD) wrote of mineralogy in similar terms of Aristotle's metaphysical theory, as the latter wrote in his pharmaceutical treatise *Bìncl̂o Gângmù* (*Compendium of Materia Medica,* 1596) (Wade-Giles spelling):

> Stone is the kernal of the chhi and the bone of the earth. In large masses it forms rocks and cliffs, in small particles it forms sand and dust. Its seminal essence (ching) becomes gold and jade; its poisonous principle becomes arsenolite (yu) and arsenious acid (phi). When the chhi becomes congealed (ning) it forms cinnabar and green vitriol. When the chhi undergoes transformation, it becomes liquid and gives rise to alums and to mercury. Its changes (are manifold), for that which is soft can become hard, as in the case of milky brine which sets to rock (-salt); and that which moves can become immobile, as in the case of the petrification of herbs, trees, or even of flying or creeping animals, which once had animation yet turn into that which has it not. Again, when thunder or thunderbolts turn to stones, there is a transformation of the formless into that which has form.

Theories of Shen Kuo

The medieval Chinese Song Dynasty statesman and scientist Shen Kuo (1031-1095 AD) wrote of his land formation theory involving concepts of mineralogy. In his *Meng Xi Bi Tan,* Shen formulated a hypothesis for the process of land formation (geomorphology); based on his observation of fossil shells in a geological stratum in a mountain hundreds of miles from the ocean. He inferred that the land was formed by erosion of the mountains and by deposition of silt, and described soil erosion, sedimentation and uplift. In an earlier work of his (circa 1080), he wrote of a curious fossil of a sea-orientated creature found far inland (Wade-Giles spelling):

> In the Chih-Phing reign period (1064 to 1067) a man of Tsechow was digging a well in his garden, and unearthed something shaped like a squirming serpent, or dragon. He was so frightened by it that he dared not touch it, but after some time, seeing that it did not move, he examined it and found it to be stone. The ignorant country people smashed it, but Chheng Po-Shun, who was magistrate of Chin-chheng at the time, got hold of a large piece of it on which scale-like markings were to be seen exactly like those on a living creature. Thus a serpent or some kind of marine snake (chhen) had certainly been turned to stone, as happens with the 'stone-crabs'.

It is of interest to note that the contemporary author of the *Xi Chi Cong Yu* attributed the idea of particular places under the sea where serpents and crabs were petrified to one Wang Jin-chen. With Shen Kuo's writing, he continued on the discovery of fossils, and his ideas of shifting geographical climates throughout time (Wade-Giles spelling):

> In recent years [circa +1080] there was a landslide on the bank of a large river in Yung-Ning Kuan near Yenchow. The bank collapsed, opening a space of several dozens of feet, and under the ground a forest of bamboo shoots was thus revealed. It contained several hundred bamboo with their roots and trunks all complete, and all turned to stone. A high official happened to pass by, and took away several, saying that he would present them to the

> emperor. Now bamboos do not grow in Yenchow. These were several dozens of feet below the present surface of the ground, and we do not know in what dynasty they could possibly have grown. Perhaps in very ancient times the climate was different so that the place was low, damp, gloomy, and suitable for bamboos. On the Chin-Hua Shan in Wuchow there are stone pine-cones, and stones formed from peach kernals, stone bulrush roots, stone fishes, crabs, and so on, but as these are all (modern) native products of that place, people are not very surprised at them. But these petrified bamboos appeared under the ground so deep, though they are not produced in that place today. This is a very strange thing.

In a similar way, the historian Joseph Needham likened Shen's account with the Scottish scientist Roderick Murchison (1792-1871), who was inspired to become a geologist after observing a providential landslide.

Modern Mineralogy

Historically, mineralogy was heavily concerned with taxonomy of the rock-forming minerals; to this end, the International Mineralogical Association is an organisation whose members represent mineralogists in individual countries. Its activities include managing the naming of minerals (via the Commission of New Minerals and Mineral Names), location of known minerals, etc. As of 2004 there are over 4000 species of mineral recognised by the IMA. Of these, perhaps 150 can be called "common," another 50 are "occasional," and the rest are "rare" to "extremely rare."

More recently, driven by advances in experimental technique (such as neutron diffraction) and available computational power, the latter of which has enabled extremely accurate atomic-scale simulations of the behaviour of crystals, the science has branched out to consider more general problems in the fields of inorganic chemistry and solid-state physics. It, however, retains a focus on the crystal structures commonly encountered in rock-forming minerals (such as the perovskites, clay minerals and framework silicates).

In particular, the field has made great advances in the understanding of the relationship between the atomic-scale structure of minerals and their function; in nature, prominent examples would be accurate measurement and prediction of the elastic properties of minerals, which has led to new insight into seismological behaviour of rocks and depth-related discontinuities in seismograms of the Earth's mantle. To this end, in their focus on the connection between atomic-scale phenomena and macroscopic properties, the mineral sciences (as they are now commonly known) display perhaps more of an overlap with materials science than any other discipline.

Physical Mineralogy

Physical mineralogy is the specific focus on physical attributes of minerals. Description of physical attributes is the simplest way to identify, classify, and categorise minerals, and they include:

- Crystallography
- Cleavage
- Lustre
- Colour
- Streak
- Hardness
- Specific gravity

Chemical Mineralogy

Chemical mineralogy focuses on the chemical composition of minerals in order to identify, classify, and categorise them, as well as a means to find beneficial uses from them. There are a few minerals which are classified as whole elements, including sulphur, copper, silver, and gold, yet the vast majority of minerals are comprised of chemical compounds, some more complex than others.

In terms of major chemical divisions of minerals, most are placed within the isomorphous groups, which are based on analogous chemical composition and similar crystal forms. A good example of isomorphism classification would be the calcite group, containing the minerals calcite, magnesite, siderite, rhodocrhosite, and smithsonite.

Optical Mineralogy

Optical mineralogy is a specific focus of mineralogy that applies sources of light as a means to identify and classify minerals. All minerals which are not part of the cubic system are double refracting, where ordinary light passing through them is broken up into two plane polarised rays that travel at different velocities and refracted at different angles.

Mineral substances belonging to the cubic system pertain only one index of refraction. Hexagonal and tetragonal mineral substances have two indices, while orthorhombric, monoclinic, and triclinic substances have three indices of refraction. With opaque ore minerals, reflected light from a microscope is needed for identification.

Crystal Structure

The use of X-ray to determine the atomic arrangement of minerals is also another way to identify and classify minerals. With minerals pertaining highly complex compositions, the exact formula of the mineral's composition can be easily discerned with knowledge of its structure. The structure of a mineral also offers a precise way of establishing isomorphism. With crystal structure, one may also deduce the correlation between atomic positions and specific physical properties.

Formation and Occurrence

The effects provided by variables and catalysts such as pressure, temperature, and time allow for the process of the formation of minerals. This process can range from simple processes found in nature, to complex formations that take years or even centuries of time. The origin of certain minerals are certainly obvious, with those such as rock salt and gypsum from evaporating sea water. Various possible methods of formation include:

- Sublimation: formation from gases
- Formation from liquids
- Formation from metamorphism
- Formation from occurrences of other minerals

Uses of Minerals

Minerals are essential to various needs within human society, such as minerals used for bettering health and fitness (such as mineral water or commercially-sold vitamins), essential components of metal products used in various commodities and machinery, essential components to building materials such as limestone, marble, granite, gravel, glass, plaster, cement, plastics, etc. Minerals are also used in fertilizers to enrich the growth of agricultural crops.

Descriptive Mineralogy

Descriptive mineralogy summarises results of studies performed on mineral substances. It is the scholarly and scientific method of recording the identification, classification, and categorisation of minerals, their properties, and their uses. Classifications for descriptive mineralogy follow as such:

- Elements
- Sulphides and sulpho minerals
- Oxides and hydrated oxides
- Haloids

- Nitrates, carbonates, and borates
- Sulphates, chromates, molybdates, and tungstates
- Phosphates, arsenates, and vanadates
- Silicates

Determinative Mineralogy

Determinative mineralogy is the actual scientific process of identifying minerals, through data gathering and conclusion. When new minerals are discovered, a standard procedure of scientific analysis is followed, including measures to identify a mineral's formula, its crystallographic data, its optical data, as well as the general physical attributes determined and listed.

Physical Characteristics of Minerals

The physical properties of a mineral are determined by its chemical composition and its crystalline structure. Within the limits of the permissible variation in chemical composition, different samples of a single mineral species are expected to display the same set of physical properties. These characteristic physical properties are therefore very useful to the field geologist in identifying and describing a specimen.

Properties which describe the physical appearance of a mineral specimen include colour, streak, and lustre. Mass-dependent properties include density; mechanical properties include hardness, cleavage, fracture, and tenacity. Properties relating to the growth patterns and physical appearance of crystals, both individually and in aggregate, are described in terms of crystal habit, crystal form, and crystal system.

Some minerals derive their colour from the presence of an element which is defined to be part of the crystal lattice. Colouration of this type is termed allochromatic colouration. Different samples of a mineral species which possesses allochromatic colouration will all display the same colour. Other minerals are coloured by the presence of certain elements in mixture. These minerals may exhibit a range of colours. Still other mineral species may usually be colourless, but may display several startling and different colours when trace amounts of impurities are present.

This type of colouration is termed idiochromatism. Certain elements are strong pigmenting agents and may lend vivid colours to specimens when they are present, whether as a part of the crystal lattice, in mixture, or as an impurity. Such elements are called chromophores. The colour which a mineral displays when it has been ground to a fine powder is termed streak. Trace amounts of impurities do not tend to affect the streak of a mineral, so this characteristic is usually more predictable than colour.

Minerals are either opaque or transparent. A thin section of an opaque mineral will not transmit light, whereas a thin section of a transparent mineral will. Relative differences in opacity and transparency are described as lustre. The characteristic of lustre refers to the amount and quality of light which is reflected from a mineral's exterior surfaces, and provides a description of how much the mineral surface 'sparkles'.

The property of density is defined as mass per unit volume. Hardness is defined as the amount of difficulty with which a mineral specimen may be scratched. Hardness has traditionally been measured according to the Mohs scale, but the diamond indentation method now provides a more quantitative measurement. The cleavage of a crystal refers to its propensity for splitting along a smooth plane. A cleavage plane is a plane of structural weakness. The property which is termed cleavage refers to both the ease with which the mineral cleaves and to the character of the exposed surface. Not every mineral exhibits cleavage. Fracture takes place when a mineral sample is split in a direction which does not serve as a plane of cleavage. A mineral fractures when it is broken or crushed. In some mineral species fractured surfaces may possess a characteristic appearance.

The term crystal habit describes the favoured growth

pattern of the crystals of a mineral species. The crystals of particular species sometimes form very distinctive, characteristic shapes. Crystal habit is also greatly determined by the environmental conditions under which a crystal develops.

Classification of Crystals

The descriptive terminology of the discipline of crystallography is applied to crystals in order to describe their structure, symmetry, and shape. This terminology describes the crystal lattice, which provides a mineral with its ordered internal structure. It also describes and analyses various types of symmetry. By considering what type of symmetry a mineral species possesses, the species may be categorised as a member of one of six crystal systems and one of thirty-two crystal classes.

The concept of symmetry describes the periodic repetition of structural features. Two general types of symmetry exist. These include translational symmetry and point symmetry. Translational symmetry describes the periodic repetition of a motif across a length or through an area or volume. Point symmetry, on the other hand, describes the periodic repetition of a motif about a single point. Reflection, rotation, inversion, and rotoinversion are all point symmetry operations.

A specified motif which is translated linearly and repeated many times will produce a lattice. A lattice is an array of points which define a repeated spatial entity called a unit cell. The unit cell of a lattice is the smallest unit which can be repeated in three dimensions in order to construct the lattice.

The number of possible lattices is limited. In the plane only five different lattices may be produced by translation. The French crystallographer Auguste Bravais (1811-1863) established that in three-dimensional space only fourteen different lattices may be constructed. These fourteen different lattices are thus termed the Bravais lattices.

The reflection, rotation, inversion, and rotoinversion symmetry operations may be combined in a variety of different ways. There are thirty-two possible unique combinations of symmetry operations. Minerals possessing the different combinations are therefore categorised as members of thirty-two crystal classes. In this classificatory scheme each crystal class corresponds to a unique set of symmetry operations. Each of the crystal classes is named according to the variant of a crystal form which it displays. Each crystal class is grouped as one of the six different crystal systems according to which characteristic symmetry operation it possesses.

A crystal form is a set of planar faces which are geometrically equivalent and whose spatial positions are related to one another by a specified set of symmetry operations. If one face of a crystal form is defined, the specified set of point symmetry operations will determine all of the other faces of the crystal form. A simple crystal may consist of only a single crystal form. A more complicated crystal may be a combination of several different forms. Example of crystal forms are the parallelohedron, prism, pyramid, trapezohedron, rhombohedron and tetrahedron.

Each crystal class is a member of one of six crystal systems. These include the isometric, hexagonal, tetragonal, orthorhombic, monoclinic, and triclinic crystal systems. Every crystal of a certain crystal system shares a characteristic symmetry element - for example, a certain axis of rotational symmetry - with the other members of its system. The crystal system of a mineral species may sometimes be determined by examining a particularly well-formed crystal of the species.

Mineral Classification

The Berzelian mineral classification system was named in honour of the Swedish chemist and mineralogist Jons Jakob Berzelius (1779-1848). The Berzelian system categorises mineral species according to the main anion group present in their chemical composition. Mineral classes may then be further subdivided according to physical features, the presence of certain cations in the lattice, the presence or absence of water or the hydroxyl anion, or structural considerations. The main classes which are recognised under Berzelius' scheme include the native elements; sulphides and sulphosalts; oxides and hydroxides; halides; carbonates, nitrates and borates; sulphates; phosphates; and silicates. The antimonides, arsenides, selenides, and tellurides closely resemble the sulphides, while the chromates, molybdates and tungstates resemble the sulphates. The arsenates and vanadates are closely akin to the phosphates.

The native elements include all mineral species which are composed entirely of atoms in an uncombined state. Such minerals either contain the atoms of only one element or else are metal alloys. The native elements are divided into metallic, semi-metallic, and non-metallic subgroups. Metals tend to be dense and malleable substances which possess a characteristic metallic lustre and conduct electricity well, while semi-metals and non-metals tend to be brittle and to conduct poorly.

Minerals of the sulphide class are compounds which contain the non-metallic element sulphur in combination with atoms of a metal or a semi-metal. Compounds in which anions of antimony (Sb), arsenic (As), selenium (Se), or tellurium (Te) replace the sulphur anion and bond with metallic or semi-metallic cations are classed respectively as antimonides, arsenides, selenides, and tellurides. If the sulphur anion, a metallic element, and a semi-metal are all present then the mineral is categorised as one of the rare sulphosalts. Most sulphides and sulphosalts are soft, dark, heavy, and brittle, possessing a metallic lustre and high conductivity.

The minerals of the oxide class contain oxygen bonded to one or more metallic elements. Hydroxides are compounds of a metallic element and water or the hydroxyl anion (OH^-). The oxide minerals tend to be relatively hard. Many provide important metal ores, and some of them may be used as gemstones. Minerals of the hydroxide class tend to be softer and less dense than oxides.

In members of the halide class an element of the halogen group such as fluorine (F), chlorine (Cl), bromine (Br), or iodine (I) bonds to a metal or semi-metal cation such as sodium (Na), potassium (K), magnesium (Mg), calcium (Ca), aluminium (Al), copper (Cu), or silver (Ag). Halides are constructed entirely of ionic bonds. The halides tend to be soft, brittle, easily soluble in water, and possess medium to high melting points. In solid state they are poor conductors.

Mineral species which are members of the carbonate class are compounds of a metal or semi-metal with the carbonate anion $(CO_3)^{2-}$. The nitrates are structurally very closely akin to the carbonates with the nitrate anion $(NO_3)^-$ replacing the carbonate anion. The nitrates tend to be softer and possess lower melting points than do the carbonates. Minerals of the borate class contain the borate radical $(BO_3)^{3-}$ rather than the carbonate or nitrate radicals. However, borate radicals may unlike the carbonate or nitrate radicals be linked into polymerised chains, sheets, or groups. This trait grants the minerals of the borate class unique structural characteristics.

This sulphate radical $(SO_4)^{2-}$ forms the basic structural unit of the minerals of the sulphate class. Minerals of the chromate class are compounds of metallic cations with the chromate anion group $(CrO_4)^{2-}$. Just as sulphur and chromium form the anion groups $(SO_4)^{2-}$ and $(CrO_4)^{2-}$, the ions of molybdenum (Mo) and tungsten (W) bond with oxygen atoms to create the anion groups $(MoO_4)^{2-}$ and $(WO_4)^{2-}$. These anion groups then bond with metal cations to form the minerals of the molybdate and tungstate classes.

Like sulphur, the elements phosphorous (P), arsenic (As), and vanadium (V) form anion groups in combination with oxygen. The resulting phosphate radical, $(PO_4)^{3-}$, provides the basic structural unit of the minerals of the phosphate class; the arsenate and vanadate radicals $(AsO_4)^{3-}$ and $(VaO_4)^{3-}$ form the basic structural units of the arsenate and vanadate classes. The mineral species of these three classes are thus composed of the respective phosphate, arsenate, and vanadate radicals linked by various metal and semi-metal cations. The basic constituent of the minerals of the silicate class is the silicate radical $(SiO_4)^4$. Silicate radicals may remain structurally isolate, join together in pairs, or polymerise into frameworks, sheets, chains, or rings. The various species of the silicate class are grouped according to their structural type. Silicate minerals are usually of relatively great hardness, and single crystals are often translucent.

The Petrogenesis

Specimens of Collection: The small collection of minerals with which this paper is concerned contains five species. All samples in the collection were removed from the copper mines of Butte, Montanna during the 1950's. Native copper, the copper carbonates azurite and malachite, and the amorphous copper silicate chrysocolla are copper-bearing substances. Rhodocrosite ($MnCO_2$) is, like azurite and malachite, a carbonate mineral, but is an ore of manganese rather than a copper ore.

Copper (Cu) is a native metal which has long been utilised as a resource by humans. It is composed of only elemental copper and its physical features are derived from the prevalent metallic bonds of its structure. Like all metals, native copper is dense, soft, malleable, and ductile; it conducts well. It is opaque and possesses the shiny, highly reflective lustre typical of metals. In colour it is the familiar, rosy copper red; it possesses a red streak.

Native copper demonstrates a face-centred cubic lattice, which is the most dense and symmetric of all possible crystalline structures. It is of the isometric or cubic crystal system. Individual copper crystals may be of cubic, dodecahedral, tetrahedral, and very rarely of octahedral shape. Aggregates of crystals may be wiry in habit or form a dendritic, arborescent branching structure.

Rhodocrosite ($MnCO_3$) is member of the carbonate class, a member of calcite group, and an ore of manganese. Its characteristic rose pink or red colour is idiochromatic and is due to the presence of the chromophore manganese (*Mn*). Rhodocrosite is translucent and will transmit light diffusely; it possesses a vitreous or pearly lustre.

Rhodocrosite is like all calcite minerals a member of the rhombohedral crystal system. Rhodocrosite's cleavage is perfect in three directions and is rhombohedral. Crystals are rare but may be rhombohedral in shape; more typically the mineral demonstrates botryoidal, encrusting habit.

The hydrous silicate chrysocolla ($Cu_2H_2Si_2O_5(OH)_4$) is an ore of copper. Chrysocolla is not a true mineral and does not possess a crystalline lattice. Instead, it is an amorphous 'silica gel' or gelatinous precipitate. It is, however, a copper-bearing solid substance and is commonly found in association with minerals such as azurite, malachite, and native copper. Chrysocolla is green to sky-blue in colour.

This colouration is idiochromatic and is due to the presence of the chromophore copper (Cu). Chrysocolla is translucent; samples may be of vitreous or glassy lustre or appear greasy, dull, or earthy. It has no cleavage and demonstrates uneven or conchoidal fracture. Specimens of chrysocolla are typically of massive or earthy habit. The substance may also display reniform or bulbous, botryoidal habit.

2

Basic Concepts

Index Minerals

An index mineral is used in geology to determine the degree of metamorphism a rock has experienced.

Depending on the original composition of and the pressure and temperature experienced by the protolith, chemical reactions between minerals in the solid state produce new minerals.

When an index mineral is found in a metamorphosed rock, it indicates the minimum pressure and temperature the protolith must have achieved in order for that mineral to form.

The higher the pressure and temperature in which the rock formed, the higher the grade of the rock.

Examples of Index Minerals -Low Grade to High Grade:

Zeolite, Chlorite, Muscovite, Biotite, Garnet, Staurolite, Kyanite, and Sillimanite Mudrock, a fine-grained sedimentary rock often containing aluminium-rich minerals, produces these minerals after being metamorphosed.

Muscovite

Muscovite (also known as Common mica, Isenglass, or Potash mica is a phyllosilicate mineral of aluminium and potassium with formula: $KAl_2(AlSi_3O_{10})(F,OH)_2$. It has a highly perfect basal cleavage yielding remarkably thin laminae (sheets), which are often highly elastic.

Muscovite melts at approximately 1320°C, has a Mohs hardness of 2 - 2.25 and a specific gravity of 2.76 - 3. It can be colourless or tinted through grays, browns, greens, yellows, or (rarely) violet or red, and can be transparent or translucent. The green chromium rich variety is called fuchsite.

Muscovite is the most common mica, found in granites, pegmatites, gneisses and schists, and as a contact metamorphic rock or as a secondary mineral resulting from the alteration of topaz, feldspar, kyanite, etc. In pegmatites, it is often found in immense sheets that are commercially valuable. Muscovite is in demand for the manufacture of fireproofing and insulating materials and to some extent as a lubricant.

The name of muscovite comes from Muscovy-glass, a name formerly used for the mineral because of its use in Russia for windows. It is anisotropic, and has quite a high birefringence. Its crystal system is monoclinic.

Biotite

Biotite is a common phyllosilicate mineral within the mica group, with the approximate chemical formula $K(Mg, Fe)_3AlSi_3O_{10}(F, OH)_2$. More generally, it refers to the dark mica series, primarily a solid-solution series between the iron-end-member annite, and the magnesium-end-member phlogopite; more aluminous end-members include siderophyllite.

Biotite is a sheet silicate. Iron, magnesium, aluminium, silicon, oxygen, and hydrogen form sheets that are weakly bond together by potassium ions. It is sometimes called "iron mica" because it is more iron-rich than phlogopite. It is also sometimes called "black mica" as opposed to "white mica" (muscovite) — both form in some rocks, in some instances side-by-side.

Like other mica minerals, biotite has a highly perfect basal cleavage, and consists of flexible sheets, or lamellae, which easily flake off. It has a monoclinic crystal system, with tabular to prismatic crystals with an obvious pinacoid termination. It has four prism faces and two pinacoid faces to form a pseudohexagonal crystal. Although not easily seen because of the cleavage and sheets, fracture is uneven.

It has a hardness of 2.5 - 3, a specific gravity of 2.7 - 3.1, and an average density of 3.09 g/cm^3. It is coloured greenish to brown or black, and even yellow when weathered. It can be transparent to opaque, has a vitreous to pearly lustre, and a grey-white streak. In its weathered yellow, sparkly form, it is a common type of "fool's Gold" (Pyrite is the official "fool's Gold"). When biotite is found in large chunks, they are called "books" because it resembles a book with pages of many sheets.

Biotite is found in a wide variety of igneous rocks and metamorphic rocks. For instance, biotite occurs in the lava of Mount Vesuvius and at Monzoni. It is an essential phenocryst in some varieties of lamprophyre.

Biotite is occasionally found in large sheets, especially in pegmatite veins, as in New England, Virginia and North Carolina. Other notable occurrences include Bancroft and Sudbury, Ontario. It is an essential constituent of many metamorphic schists, and it forms in suitable compositions over a wide range of pressure and temperature.

It is not industrially useful, but it is mined using quarrying and underground mining (depending on the depth of the biotite) for collection purposes.

Biotite is used extensively to constrain ages of rocks, by either potassium-argon dating or argon-argon dating. Because argon escapes readily from the biotite crystal structure at high temperatures, these methods may provide only minimum ages for many rocks. Biotite is also useful in assessing temperature histories of metamorphic rocks, because the partitioning of iron and magnesium between biotite and garnet is sensitive to temperature.

Biotite is used in electrical devices. It was named by J.F.L. Hausmann in 1847 in honour of the French physicist Jean-Baptiste Biot, who, in 1816, discovered optical properties of micas.

GARNET

Garnet is a group of minerals that have been used since the Bronze Age as gemstones and abrasives. Garnets are most often seen in red, but are available in a wide variety of colours spanning the entire spectrum. The name "garnet" comes from the Latin *granatus* ("grain"), possibly a reference to the *Punica granatum* ("pomegranate"), a plant with red seeds similar in shape, size, and colour to some garnet crystals.

Six common varieties of garnet are recognised based on their chemical composition. They are pyrope, almandine, spessartite, grossular (varieties of which are hessonite or cinnamon-stone and tsavorite), uvarovite and andradite. The garnets make up two solid solution series; 1. pyrope-almandine-spessarite and 2. uvarovite-grossular-andradite.

Appearance: Garnets are found in many colours including red, orange, yellow, green, blue, purple, brown, black, pink and colourless. The rarest of these is the blue garnet, discovered in the late 1990s in Bekily, Madagascar. It is also found in parts of the United States, Russia and Turkey.

It changes colour from blue-green in the daylight to purple in incandescent light, as a result of the relatively high amounts of vanadium (about 1 wt. per cent V_2O_3). Other varieties of colour-changing garnets exist. In daylight, their colour ranges from shades of green, beige, brown, gray, and blue, but in incandescent light, they appear a reddish or purplish/pink colour. Because of their colour changing quality, this kind of garnet is often mistaken for Alexandrite.

Garnet's light transmission properties can range from the gemstone-quality transparent specimens to the opaque varieties used for industrial purposes as abrasives. The mineral's lustre is categorised as vitreous (glass-like) or resinous (amber-like).

Crystal Structure

Garnets are nesosilicates having the general formula $X_3Y_2(SiO_4)_3$. The *X* site is usually occupied by divalent cations (Ca^{2+}, Mg^{2+}, Fe^{2+}) and the *Y* site by trivalent cations (Al^{3+}, Fe^{3+}, Cr^{3+}) in an octahedral/tetrahedral framework with $[SiO_4]^{4-}$ providing the tetrahedra.

Garnets are most often found in the dodecahedral crystal habit, but are also commonly found in the trapezohedron habit. (Note: the word "trapezohedron" as used here and in most mineral texts refers to the shape called a Deltoidal icositetrahedron in solid geometry.) They crystallise in the isometric system, having three axes that are all of equal length and perpendicular to each other. Garnets do not show cleavage, so when they fracture under stress, sharp irregular pieces are formed.

Hardness

Because the chemical composition of garnet varies, the atomic bonds in some varieties are stronger than in others. As a result, the mineral shows a range of hardness on the Mohs Scale of about 6.5 to 7.5. The harder varieties, like almandite, are often used for abrasive purposes.

Garnet Group End-members

Pyralspite garnets - Al in Y site

- Almandite: $Fe_3Al_2(SiO_4)_3$
- Pyrope: $Mg_3Al_2(SiO_4)_3$
- Spessartine: $Mn_3Al_2(SiO_4)_3$

Almandite

Almandine in gneissic rock

Almandite, sometimes called almandine, is the modern gem known as carbuncle (though originally almost any red gemstone was known by this name). The term "carbuncle" is derived from the Latin meaning "little spark." The name Almandite is a corruption of Alabanda, a region in Asia Minor where these stones were cut in ancient times.

Chemically, almandite is an iron-aluminium garnet with the formula $Fe_3Al_2(SiO_4)_3$; the deep red transparent stones are often called precious garnet and are used as gemstones (being the most common of the gem garnets).

Almandite occurs in metamorphic rocks like mica schists, associated with minerals such as staurolite, kyanite, andalusite, and others. Almandite has nicknames of Oriental garnet, almandine ruby, and carbuncle.

Pyrope

Pyrope, from the Latin *pyropos*, means similar to fire. It is red in colour and chemically a magnesium aluminium silicate with the formula $Mg_3Al_2(SiO_4)_3$, though the magnesium can be replaced in part by calcium and ferrous iron. The colour of pyrope varies from deep red to almost black. Transparent pyropes are used as gemstones.

A variety of pyrope from Macon County, North Carolina

is a violet-red shade and has been called *rhodolite,* from the Greek meaning "a rose." In chemical composition it may be considered as essentially an isomorphous mixture of pyrope and almandite, in the proportion of two parts pyrope to one part almandite. Pyrope has tradenames some of which are misnomers; *Cape ruby, Arizona ruby, California ruby, Rocky Mountain ruby,* and *Bohemian garnet* from the Czech Republic. Another intriguing find is the blue colour-change garnets from Madagascar, a pyrope spessartine mix.

The colour of these blue garnets is not like sapphire blue in subdued daylight but more reminiscent of the grayish blues and greenish blues sometimes seen in spinel. However in white LED light the colour is equal to the best corn flower blue sapphire or D block tanzanite this is due to the blue garnets ability to absorb the yellow component of the emitted light.

Pyrope is an indicator mineral for high pressure rocks. The garnets from mantle derived rocks, peridotites and eclogites, commonly contain a pyrope variety.

Spessartite

Spessartite or spessartine is manganese aluminium garnet, $Mn_3Al_2(SiO_4)_3$. Its name is derived from Spessart in Bavaria. It occurs most often in granite pegmatite and allied rock types and in certain low grade metamorphic phyllites. Spessartite of a beautiful orange-yellow is found in Madagascar. Violet-red spessartites are found in rhyolites in Colorado and Maine.

Ugrandite Group - Calcium in X Site

- Andradite: $Ca_3Fe_2(SiO_4)_3$
- Grossular: $Ca_3Al_2(SiO_4)_3$
- Uvarovite: $Ca_3Cr_2(SiO_4)_3$

Andradite

Andradite is a calcium-iron garnet, $Ca_3Fe_2(SiO_4)_3$, is of variable composition and may be red, yellow, brown, green or black. The recognised subvarieties are topazolite (yellow or green), demantoid (green) and melantite (black). Andradite is found both in deep-seated igneous rocks like syenite as well as serpentines, schists, and crystalline limestone. Demantoid has been called the "emerald of the Urals" from its occurrence there, and is one of the most prized of garnet varieties. Topazolite is a golden yellow variety and melanite is a black variety.

Grossular

Grossular is a calcium-aluminium garnet with the formula $Ca_3Al_2(SiO_4)_3$, though the calcium may in part be replaced by ferrous iron and the aluminium by ferric iron. The name grossular is derived from the botanical name for the gooseberry, *grossularia*, in reference to the green garnet of this composition that is found in Siberia. Other shades include cinnamon brown (cinnamon stone variety), red, and yellow. Because of its inferior hardness to zircon, which the yellow crystals resemble, they have also been called *hessonite* from the Greek meaning inferior. Grossular is found in contact metamorphosed limestones with vesuvianite, diopside, wollastonite and wernerite.

One of the most sought after varieties of gem garnet is the fine green grossular garnet from Kenya and Tanzania called tsavorite. This garnet was discovered in the 1960s in the Tsavo area of Kenya, from which the gem takes its name.

Uvarovite

Uvarovite is a calcium chromium garnet with the formula $Ca_3Cr_2(SiO_4)_3$. This is a rather rare garnet, bright green in colour, usually found as small crystals associated with chromite in peridotite, serpentinite, and kimberlites. It is found in crystalline marbles and schists in the Ural mountains of Russia and Outokumpu, Finland.

Less common species

- Calcium in *X* site
  - — Goldmanite: $Ca_3V_2(SiO_4)_3$

- — Kimzeyite: $Ca_3(Zr,Ti)_2[(Si,Al,Fe^{3+})O_4]_3$
- — Morimotoite: $Ca_3Ti^{4+}Fe^{2+}(SiO_4)_3$
- — Schorlomite: $Ca_3(Ti^{4+},Fe^{3+})_2[(Si,Ti)O_4]_3$
- Hydroxide bearing - calcium in *X* site
- Hydrogrossular: $Ca_3Al_2(SiO_4)_{3-x}(OH)_{4x}$
  - — Hibschite: $Ca_3Al_2(SiO_4)_{3-x}(OH)_{4x}$ (where x is between 0.2 and 1.5)
  - — Katoite: $Ca_3Al_2(SiO_4)_{3-x}(OH)_{4x}$ (where x is greater than 1.5)
- Magnesium or manganese in *X* site
  - — Knorringite: $Mg_3Cr_2(SiO_4)_3$
  - — Majorite: $Mg_3(Fe,Al,Si)_2(SiO_4)_3$
- Calderite: $Mn_3Fe^{3+}{}_2(SiO_4)_3$

Knorringite

Knorringite is a magnesium chromium garnet species with the formula $Mg_3Cr_2(SiO_4)_3$. Pure end-member knorringite never occurs in nature.

Knorringite is only formed under high pressure and is often found in kimberlites. It is used as an indicator mineral in the search for diamonds.

Synthetic Garnets

In yttrium iron garnet (YIG), $Y_3Fe_2(FeO_4)_3$, the five iron (III) ions occupy two octahedral and three tetrahedral sites, with the yttrium (III) ions coordinated by eight oxygen ions in an irregular cube. The iron ions in the two coordination sites exhibit different spins, resulting in magnetic behaviour. YIG is a ferromagnetic material having a Curie temperature of 550 K. By substituting specific sites with rare earth elements, for example, interesting magnetic properties can be obtained.

One example for this is gadolinium gallium garnet, $Gd_3Ga_2(GaO_4)_3$, which is synthesised for use in magnetic bubble memory. Yttrium aluminium garnet (YAG), $Y_3Al_2(AlO_4)_3$, is used for synthetic gemstone. When doped with neodymium (Nd^{3+}), these YAl-Garnets are useful as the lasing medium in lasers.

Uses of Garnets: Pure crystals of garnet are used as gemstones. In particular, it is the birthstone for the month of January. Garnet sand is a good abrasive, and a common replacement for silica sand in sand blasting. Mixed with very high pressure water, garnet is used to cut steel and other materials in water jets.

Garnet is a key mineral in interpreting the genesis of many igneous and metamorphic rocks via geothermobarometry. Diffusion of elements is relatively slow in garnet compared to rates in many other minerals, and garnets are also relatively resistant to alteration. Hence, individual garnets commonly preserve compositional zonations that are used to interpret the temperature-time histories of the rocks in which they grew. Garnet grains that lack compositional zonation commonly are interpreted as having been homogenised by diffusion, and the inferred homogenisation also has implications for the temperature-time history of the host rock.

Garnets are also useful in defining metamorphic facies of rocks. For instance, eclogite can be defined as a rock of basalt composition, but mainly consisting of garnet and omphacite. Pyrope-rich garnet is restricted to relatively high-pressure metamorphic rocks, such as those in the lower crust and in the Earth's mantle.

Peridotite may contain plagioclase, or aluminium-rich spinel, or pyrope-rich garnet, and the presence of each of the three minerals defines a pressure-temperature range in which the mineral could equilibrate with olivine plus pyroxene: the three are listed in order of increasing pressure for stability of the peridotite mineral assemblage.

Hence, garnet peridotite must have been formed at great depth in the earth. Xenoliths of garnet peridotite have been carried up from depths of 100 km and greater by kimberlite, and garnets from such disaggegated xenoliths are used as a kimberlite indicator minerals in diamond prospecting. At depths of about 300 to 400 km and greater, a pyroxene component is dissolved in garnet, by the substitution of (Mg,Fe) plus Si for 2Al in the octahedral (Y) site in the garnet structure, creating unusually silica-rich garnets that have solid solution towards majorite. Such silica-rich garnets have been identified as inclusions within diamonds.

In the USA it is known as the birthstone for January. It has been made the state mineral of Connecticut.

Staurolite

Staurolite is a red brown to black, mostly opaque, nesosilicate mineral with a white streak. It crystallises in the monoclinic crystal system, has a Mohs hardness of 7 to 7.5 and a rather complex chemical formula: $(Fe,Mg,Zn)_2Al_9(Si,Al)_4O_{22}(OH)_2$. Iron, magnesium and zinc occur in variable ratios. Staurolite is mostly opaque.

A special property of staurolite is that it often occurs twinned in a characteristic cross-shape. The macroscopically visible crystals are of prismatic shape. They are often larger than the surrounding minerals and are then called porphyroblasts.

The name is derived from the Greek, *stauros* for cross and lithos for stone in reference to the common twinning. Staurolite is a regional metamorphic mineral of intermediate to high grade. It occurs with almandine garnet, micas, kyanite and other metamorphic minerals. It is the official state mineral of Georgia.

Kyanite

Kyanite, whose name derives from the Greek word *kyanos*, meaning blue, is a typically blue silicate mineral, commonly found in aluminium-rich metamorphic pegmatites and/or sedimentary rock. Kyanite is a diagnostic mineral of the Blueschist Facies of metamorphic rocks.

Kyanite is a member of the aluminosilicate series, which includes the polymorph andalusite and the polymorph sillimanite. Kyanite is strongly anisotropic, in that its hardness varies depending on its crystallographic direction. While this is a feature of almost all minerals, in kyanite this anisotropism can be considered an identifying characteristic.

Uses of Kyanite

Kyanite is used primarily in refractory and ceramic products, including porcelain plumbing fixtures and dinnerware. It is also used in electrical insulators and abrasives. An interesting property of kyanite is that it undergoes an irreversible expansion when fired at high temperature. Kyanite has also been used as a gemstone, though this use is limited by its anisotropism and perfect cleavage. Finally, as with most minerals, kyanite is a collector's mineral.

Associated Minerals

Kyanite is usually found in association with its polymorphs, as well as other silicate minerals.

These include:

- andalusite, Al_2SiO_5
- sillimanite, Al_2SiO_5
- quartz, SiO_2
- staurolite, $Fe_2Al_9Si_4O_{22}(OH)_2$
- micas, $AB_{2\text{-}3}(X, Si)_4O_{10}(O,F,OH)_2$
- garnets, $A_3B_2(SiO_4)_3$

Alternative Names

Kyanite has several alternative names, including disthene, munkrudite and cyanite. White-grey kyanite is also called rhaeticite.

Notes for Identification: Kyanite's elongated, columnar crystals are usually a good first indication of the mineral, as well as its colour (when the specimen is blue). Associated minerals are useful as well, especially the presence of the polymorphs or staurolite, which occur frequently with kyanite. However, the most useful characteristic in identifying kyanite is its anisotropism. If one suspects a specimen to be kyanite, verifying that it has two distinctly different hardnesses on perpendicular axes is a key to identification.

SILLIMANITE

Sillimanite also called Bucholzite is an alumino-silicate mineral with the chemical formula Al_2SiO_5. Sillimanite is named after the American chemist Benjamin Silliman (1779-1864).

Occurrence

Sillimanite is one of three alumino-silicate polymorphs, the other two being andalusite and kyanite. A common variety of sillimanite is known as fibrolite, so named because the mineral appears like a bunch of fibres twisted together when viewed under thin section or even by the naked eye. Both the fibrous and traditional forms of sillimanite are common in metamorphosed sedimentary rocks. It is an index mineral indicating a high degree of metamorphism.

Crystallographic system	*Orthorhombic*
Colours	White, brown and green
Lustre	Silky if in fibrous form, else vitreous
Hardness	7.5
Refractive index	1.641-1.648
Cleavage	1
Twinning	n/a
Pleochroism	Colourless to pale brown to yellow
Type locality	Vltava, Susice, Czech Republic

METAMICTISATION

Metamictisation (sometimes called metamiction) is a natural process resulting in the gradual and ultimately complete destruction of a mineral's crystal lattice, leaving the mineral amorphous. Affected material is therefore described as metamict.

Certain minerals occasionally contain interstitial impurities of radioactive compounds and it is the alpha radiation emitted from these compounds that is responsible for degrading a mineral's crystal structure through internal bombardment. Effects of metamictisation are extensive: other than negating any birefringence previously present, the process also lowers a mineral's refractive index, hardness, and specific gravity. The mineral's colour is also affected: metamict specimens are usually green or brown. Further, metamictisation diffuses the bands of a mineral's absorption spectrum. Curiously and inexplicably, the one attribute which metamictisation does not alter is dispersion. All metamict materials are themselves radioactive, some dangerously so.

An example of a metamict mineral is zircon. The presence of the uranium mineral coffinite and the thorium mineral thorite are responsible for the radiation damage in this case. Unaffected specimens are termed high zircon while metamict specimens are termed low zircon. Specimens falling between the two extremes are termed intermediate. Other minerals known to undergo metamictisation include ekanite and titanite. Ekanite is almost invariably found completely metamict as thorium and uranium are part of its essential chemical composition.

Metamict minerals can have their crystallinity and properties restored through prolonged annealing.

Miarolitic Cavities

Miarolitic cavities (or miarolitic texture) are typically crystal lined irregular cavities or vugs most commonly found in granitic pegmatites, but also found in a variety of igneous rocks. The central portions of pegmatites are often miarolitic as the pegmatite dike crystallises from the outside walls towards the centre. The volatile portion of the magma is gradually excluded from the forming crystal phases until it becomes trapped within the body and forms the cavities which often contain minerals of elements incompatible with the typical silicate granitic mineralogy.

The miarolitic cavities and miarolitic pegmatites are sources of rare and unusual minerals containing elements not found in abundance in normal igneous rocks. Minerals containing lithium, rubidium, beryllium, boron, niobium, tantalum, tin, bismuth, fluorine and other elements are found.

The term *miarlitic* comes from the Italian *miarole* in reference to the mineral rich pegmatite region of Baveno and Cuasso al Monte in northern Italy.

Normative Mineralogy

Normative mineralogy is a geochemical calculation of the whole rock geochemistry of a rock sample which estimates the idealised mineralogy of a rock according to the principles of geochemistry.

Normative mineral calculations can be achieved via either the CIPW Norm or the Barth-Niggli Norm (also known as the Cation Norm).

Normative calculations are used to produce an idealised mineralogy of a crystallised melt. First, a rock is chemically analysed to determine the elemental constituents. Results of the chemical analysis traditionally are expressed as oxides (e.g., weight per cent Mg is expressed as weight per cent MgO). The normative mineralogy of the rock then is calculated, based upon assumptions about the order of mineral formation and known phase relationships of rocks and minerals, and using simplified mineral formulas. The calculated mineralogy can be used to assess concepts such as silica saturation of melts.

The process blends the excitement of geochemistry with that of spreadsheet accounting. Because the normative calculation is essentially a computation, it can be achieved relatively painlessly via computer programmes.

CIPW Norm

The CIPW Norm was developed by the petrologists Cross, Iddings, Pirsson and the geochemist Washington. The CIPW normative mineralogy calculation is based on the typical minerals that may be precipitated from an anhydrous melt at low pressure, and simplifies the typical igneous geochemistry seen in nature with the following four constraints;

1. The magma crystallises under anhydrous conditions so that no hydrous minerals (hornblende, biotite) are formed.
2. The feromagnesium minerals are assumed to be free of Al_2O_3.
3. The Fe/Mg ratio for all feromagnesium minerals is assumed to be the same.
4. Several minerals are assumed to be incompatible, thus nepheline and/or olivine never appear with quartz in the norm.

This is, of course, an artificial set of constraints, and means that the results of the CIPW norm do not reflect the true course of igneous differentiation in nature.

The primary benefit of calculating a CIPW norm is determining what the ideal mineralogy of an aphanitic or porphyritic igneous rock is. Secondly, the degree of silica saturation of the melt which formed the rock can be assessed in the absence of diagnostic feldspathoid species.

The silica saturation of a rock varies not only with silica content but the proportion of the various alkalis and metal species within the melt. The silica saturation eutectic plane is thus different for various families of rocks and cannot be easily estimated, hence the requirement to calculate whether the rock is silica saturated or not.

This is achieved by assigning cations of the major elements within the rock to silica anions in modal proportion, to form solid solution minerals in the idealised mineral assemblage starting with phosphorus for apatite, chlorine and sodium for halite, sulphur and FeO into pyrite, FeO and Cr_2O_3 is allocated for chromite, FeO and equal molar amount of TiO_2 for ilmenite, CaO and CO_2 for calcite, to complete the most common non-silicate minerals. From the remaining chemical constituents, Al_2O_3 and K_2O are allocated with silica for orthoclase; sodium, aluminium and potassium for albite, and so on until either there is no silica left (in which case feldspathoids are calculated) or excess, in which case the rock contains normative quartz.

Normative and Modal Mineralogy

Normative mineralogy is an *estimate* of the mineralogy of the rock. It will usually differ from the visually observable mineralogy, at least inasmuch as the types of mineral species, especially amongst the ferromagnesian minerals and feldspars, where it is possible to have many solid solution series of minerals, or minerals with similar Fe and Mg ratios substituting, especially with water (e.g., amphibole and biotite replacing pyroxene).

However, in aphanites, or rocks with phenocrysts which are clearly out of equilibrium with the groundmass, a normative mineral calculation is often the best way of gaining an understanding of the evolution of the rock and its relationship to other igneous rocks in the region.

Cautions

The CIPW Norm or Cation Norm is a useful tool for assessing silica saturation or oversaturation; estimations of minerals in a mathematical model are based on many assumptions and the results must be balanced with the observable mineralogy. The following areas create the most errors in calculations;

- *Cumulate rocks:* While a normative mineral calculation isn't necessarily improper for use on cumulate rocks, the information gleaned is of doubtful value because a cumulate rock does not represent the melt from which it was extracted. However, if the groundmass of a cumulate can be analysed, it is valid to use a normative calculation to gain information about the parental melt.

- *Oxidation state:* Because the normative calculation divides Fe between oxide phases and availability for silicate phases, based on estimates of the ratio of Fe^{2+}/Fe^{3+}, miscalculating the ratio for the rock in question may produce erroneous amounts of magnetite or haematite, or alter the silicate mineralogy. If the Fe^{2+}/Fe^{3+} ratio is known for the sample, the resulting calculation should match the observed mineralogy more closely.
- *Pressure and temperature:* Because the CIPW Norm is based on anhydrous melts and crystallisation at fairly low pressures, the resultant normative mineralogy will not reflect observed mineralogy for all rock types, especially those which are formed within the mantle. The normative mineralogy is not entirely accurate at reflecting the mineralogy of rocks formed at high pressures where, for instance, phlogopite may substitute for amphibole, amphibole for olivine and so forth. Altered normative calculations have been developed which more correctly reflect the particular pressure regimes of the deep crust and mantle.
- *Carbon dioxide:* The influence of CO_2 on most silicate melts is fairly minor but in some cases, especially carbonatite, but also certain lamprophyre type rocks, kimberlite and lamproite, the presence of carbon dioxide and calcite within the melt or accessory phases will derive erroneous normative mineralogy. This is because if carbon is not analysed, there will be excess calcium, causing normative silica undersaturation, and increasing the calcium silicate mineral budget. Similarly, if graphite is present (as is the case with some kimberlites) this can produce excess C, and hence skew the calculation towards excess carbonate. Excess

elemental C will also, in nature, result in reduced oxygen fugacity and alter Fe^{2+}/Fe^{3+} ratios.

- *Halides:* The presence of some halides and non-metallic elements in the melt will alter the resulting chemistry. For instance, boron will form tourmaline; excess chlorine may form scapolite instead of feldspar. This is generally rare, except in some A-type granites and related rocks.
- *Mineral disequilibrium:* Similar to cumulate rocks, a rock may contain extraneous minerals inherited from earlier melts, or may even contain xenoliths or restite. It would be improper to calculate normative mineralogy on an igneous breccia, for instance.

For this reason it is not advised to utilise a CIPW norm on kimberlites, lamproites, lamprophyres and some silica-undersaturated igneous rocks. In the case of carbonatite, it is improper to use a CIPW norm upon a melt which is very rich in carbonate.

It is possible to apply the CIPW norm to metamorphosed igneous rocks. The validity of the method holds as true for metamorphosed igneous rocks as any igneous rock, and in this case it is useful in deriving an assumed mineralogy from a rock which may have no remnant protolith mineralogy remaining.

PERTHITE

Perthite is used to describe an intergrowth of two feldspars: a host grain of potassium-rich alkali feldspar (near K-feldspar, $KAlSi_3O_8$, in composition) includes exsolved lamellae or irregular intergrowths of sodic alkali feldspar (near albite, $NaAlSi_3O_8$, in composition). Typically the host grain is orthoclase or microcline, and the lamellae are albite. If sodic feldspar is the dominant phase, the result is an antiperthite.

The intergrowth forms by exsolution due to cooling of a grain of alkali feldspar with a composition intermediate between K-feldspar and albite. There is complete solid solution between albite and K-feldspar at temperatures near 700°C and pressures like those within the crust of the Earth, but a miscibility gap is present at lower temperatures. If an alkali feldspar grain with an intermediate composition cools slowly enough, domains rich in more K-rich and more Na-rich feldspar separate from one another. In the presence of water, the process occurs quickly.

When megascopically developed, the texture may consist of distinct pink and white lamellae representing exsolved white albite ($NaAlSi_3O_8$) in pink microcline K-feldspar. The intergrowths in perthite have a great variety of shapes. If cooling is sufficiently slow, the alkali feldspar may exsolve to form separate grains with near-end-member albite and K-feldspar compositions.

The gem varieties of potassium feldspar, amazonite and moonstone are variant coloured perthites.

Pleochroism

Pleochroism is an optical phenomenon in which the stone appears to be different colours when observed at different angles. Pleochroism is caused by the double refraction of light by a coloured gem or crystal. Light of different polarisations is bent different amounts by the crystal, and therefore follows different paths through the crystal.

The components of a divided light beam follow different paths within the stone and travel at different speeds, and each path will absorb different colours of light. When the stone is observed at some angle, light following some combination of paths and polarisations will be present, each of which will have had light of different colours absorbed. At another angle, the light passing through the stone will be composed of another combination of light paths and polarisations, each with their own colour. The light passing through the stone will therefore have different colours when it is viewed from different angles, making the stone seem to be of different colours.

Some stones show two colours or shades and are called dichroic. Some show three and are trichroic. Gems are sometimes cut and set either to display pleochroism or to hide it, depending on the colours and their attractiveness.

Pleochroism is an extremely useful tool in mineralogy for mineral identification, since minerals that are otherwise very similar often have very different pleochroic colour schemes. In such cases, a thin section of the mineral is used and examined under transmitted light in a microscope.

BIREFRINGENCE

Birefringence, or double refraction, is the decomposition of a ray of light into two rays (the ordinary ray and the extraordinary ray) when it passes through certain types of material, such as calcite crystals, depending on the polarisation of the light. This effect can occur only if the structure of the material is anisotropic. If the material has a single axis of anisotropy, (i.e. it is uniaxial) birefringence can be formalised by assigning two different refractive indices to the material for different polarisations. The birefringence magnitude is then defined by:

$$\Delta n = n_e - n_o$$

where n_o and n_e are the refractive indices for polarisations perpendicular (ordinary) and parallel (extraordinary) to the axis of anisotropy respectively.

Birefringence can also arise in magnetic, not dielectric, materials, but substantial variations in magnetic permeability of materials are rare at optical frequencies.

Examples of Birefringent Materials

Many plastics are birefringent, because their molecules are 'frozen' in a stretched conformation when the plastic is moulded or extruded. For example, cellophane is a cheap birefringent material. Birefringent materials are used in many devices which manipulate the polarisation of light, such as wave plates, polarising prisms, and Lyot filters. There are many birefringent crystals: birefringence was first described in calcite crystals by the Danish scientist Rasmus Bartholin in 1669.

Birefringence can be observed in amyloid plaque deposits such as are found in the brains of Alzheimer's victims. Modified proteins such as immunoglobulin light chains abnormally accumulate between cells, forming fibrils. Multiple folds of these fibres line up and take on a beta-pleated sheet conformation. Congo red dye intercalates between the folds and, when observed under polarised light, causes birefringence.

Cotton (Gossypium hirsutum) fibre is birefringent because of high levels of cellulosic material in the fibre's secondary cell wall. Slight imperfections in optical fibre can cause birefringence, which can cause distortion in fibre-optic communication.

Silicon carbide, also known as Moissanite, is strongly birefringent. The refractive indices of several (uniaxial) birefringent materials are listed below (at wavelength ~ 590 nm), from.

Material	n_o	n_e	Δn
beryl $Be_3Al_2(SiO_3)6$	1.602	1.557	-0.045
calcite $CaCO_3$	1.658	1.486	-0.172
calomel Hg_2Cl_2	1.973	2.656	+0.683
ice H_2O	1.309	1.313	+0.014
lithium niobate $LiNbO_3$	2.272	2.187	-0.085
magnesium fluoride MgF_2	1.380	1.385	+0.006
quartz SiO_2	1.544	1.553	+0.009
ruby Al_2O_3	1.770	1.762	-0.008
rutile TiO_2	2.616	2.903	+0.287
peridot (Mg, Fe)2SiO_4	1.690	1.654	-0.036
sapphire Al_2O_3	1.768	1.760	-0.008
sodium nitrate $NaNO_3$	1.587	1.336	-0.251
tourmaline (complex silicate)	1.669	1.638	-0.031
zircon, high $ZrSiO_4$	1.960	2.015	+0.055
zircon, low $ZrSiO_4$	1.920	1.967	+0.047

Biaxial Birefringence

Biaxial birefringence, also known as trirefringence, describes an anisotropic material that has more than one axis of anisotropy. For such a material, the refractive index tensor n, will in general have three distinct eigenvalues that can be labelled $n_{á}$, $n_{â}$ and $n_{ã}$.

The refractive indices of some trirefringent materials are listed below (at wavelength ~ 590 nm), from.

Material	$n_{á}$	$n_{â}$	$n_{ã}$
borax	1.447	1.469	1.472
epsom salt $MgSO_4 \cdot 7(H_2O)$	1.433	1.455	1.461
mica, biotite	1.595	1.640	1.640
mica, muscovite	1.563	1.596	1.601
olivine $(Mg, Fe)_2SiO_4$	1.640	1.660	1.680
perovskite $CaTiO_3$	2.300	2.340	2.380
topaz	1.618	1.620	1.627
ulexite	1.490	1.510	1.520

Measuring Birefringence

Birefringence and related optical effects (such as optical rotation and linear or circular dichroism) can be measured by measuring the changes in the polarisation of light passing through the material. These measurements are known as polarimetry. A common feature of optical microscopes is a pair of crossed polarising filters. Between the crossed polarisers, a birefringent sample will appear bright against a dark (isotropic) background.

Applications of Birefringence

Birefringence is widely used in optical devices, such as liquid crystal displays, light modulators, colour filters, wave plates, optical axis gratings, etc. It also plays important role in second harmonic generation and many other non-linear processes. It is also utilised in medical diagnostics. Needle biopsy of suspected gouty joints will be negatively birefringent if urate crystals are present.

Elastic Birefringence

Another form of birefringence is observed in anisotropic elastic materials. The study of birefringent shear waves in the earth is a part of seismology. Birefringence is also used in optical mineralogy to determine the chemical composition, and history of minerals and rocks.

Periodic Polling

Periodic polling is formation of layers with alternate orientation in a birefringent material. The domains are regularly spaced, with period in a multiple of the desired wavelength of operation. The structure is desired to achieve quasi-phase-matching (QPM) in the material.

Periodically polled crystals are frequently used as non-linear optical materials. They are up to 20 times more efficient at second-harmonic generation than crystals of the same material without periodic structure. The material for the crystals is usually a wide band gap inorganic crystal, or in some cases a suitable organic polymer. Some popular materials in current use are KTP, lithium niobate, and lithium tantalate.

The periodic structure is created in the crystal using a range of techniques. Pulsed electric field, electron bombardment, thermal pulsing, or other methods can be used to reposition the atoms in the lattice, creating oriented domains. This can be achieved either during the growth of the crystal, or subsequently.

Pseudomorph

In geology, a pseudomorph is a mineral compound resulting from a substitution process in which the appearance and dimensions remain constant, but the mineral which makes up the chief component of the compound is replaced by another. The name literally means False Form.

Different kinds of pseudomorph exist:

A paramorph (also called allomorph) is a

mineral changed on the molecular level only. It has the same chemical composition, but with a different structure. The mineral looks identical to the original unaltered form. This occurs, as an example, in the aragonite to calcite change.

An infiltration pseudomorph is a pseudomorph in which one mineral or other material is replaced by another. The original shape of the mineral remains unchanged, but colour, hardness, and other properties change to those of the replacing mineral. An example of this process (also called substitution) is the replacement of wood by silica (quartz or opal) to form petrified wood in which the substitution may be so perfect as to retain the original cellular structure of the wood. An example of mineral-to-mineral substitution is replacement of aragonite twin crystals by native copper, as occurs at Corococo, Bolivia.

A variety of infiltration or substitution pseudomorphism is called alteration, in which only partial replacement occurs. This happens typically when a mineral of one composition changes by chemical reaction to another of similar composition, retaining the original crystalline shape. An example is a change from galena, lead sulphide, to anglesite, lead sulphate. The resulting pseudomorph may contain an unaltered core of galena surrounded by anglesite that has the cubic crystal shape of galena.

An incrustation pseudomorph results from a process by which a mineral is coated by another and the encased mineral dissolves. The encasing mineral remains intact, and retains the shape of the original mineral or material.

Alternatively, another mineral may fill the space (the mold) previously occupied by some other mineral or material.

Pseudomorphs are also common in palaeontology. Fossils are often formed by pseudomorphic replacement of the remains by mineral matter. Examples would include petrified wood and pyritised gastropod shells.

Terminology for pseudomorphs is "replacer after original", as in brookite after rutile.

Pyrognomic

Pyrognomic materials are easily made incandescent. Minerals and metals commonly glow when heated, but those that are pyrognomic glow at much lower temperatures. Allanite and gadolinite are examples of pyrognomic minerals. The term was introduced by Scheerer in 1840, but the phenomenon had been previously observed by Wollaston and Berzelius.

Vitreous

Vitreous refers to a material in an amorphous, glassy state (in contrast to a crystalline state). In such a state, the constituent atoms do not exhibit the long-range order that is characteristic of crystals. However, they still exhibit short-range order — the separation of atoms and/or the lengths of covalent bonds are very close to their typical equilibrium distances.

The creation of a vitreous material by supercooling or by addition of additives prior to cooling is called vitrification. Toilets are made of vitreous china, one of the few materials able to withstand the harsh (caustic) chemicals used to clean them.

Under certain conditions, vitreous materials may begin to rearrange themselves into crystals. This process is referred to as devitrification. Natural volcanic glass, obsidian will de-vitrify or crystallise over extended periods of geologic time, which can also occur much more quickly in dishwashers due to the chemicals involved in their detergents.

Other Uses:

- When applied to whiteware ceramics vitreous means the material has an extremely low permeability to liquids, often but not always water, when determined by a specified test regime. The microstructure of whiteware ceramics frequently contain both amorphous and crystalline phases.
- A vitreous lustre in mineralogy refers to a glassy lustre or sheen of a mineral surface. The minerals quartz and fluorite exhibit a vitreous lustre.

WIGNER-SEITZ CELL

The Wigner-Seitz cell is a geometrical construction which helps in the study of crystalline material in solid-state physics. The unique property of a crystal is that the atoms comprising it are arranged in a regular, 3-dimensional array, which is called a lattice. All the properties attributed to crystalline materials stem from this highly ordered structure. Such a structure exhibits discrete translational symmetry. In order to model and study such a periodic system, one needs a mathematical "handle" to describe the symmetry and hence draw conclusions about the consequences of this symmetry. The Wigner-Seitz cell is a means to achieve this.

Definition

The Wigner-Seitz cell around a lattice point is defined as the locus of points in space, which are closer to that lattice point than to any of the other lattice points. It can be shown mathematically that a Wigner-Seitz Cell is a primitive unit cell spanning the entire Bravais lattice without leaving any gaps or holes.

Constructing the Cell: The cell may be chosen by first picking a lattice point. Then, lines are drawn to all nearby (closest) lattice points. At the midpoint of each line, another line is drawn normal to each of the first set of lines. In the case of a three-dimensional lattice, a perpendicular plane is drawn at the midpoint of the lines between the lattice points. By using this method, the smallest area (or volume) is enclosed in this way and is called the Wigner-Seitz primitive cell. All area (or space) within the lattice will be filled by this type of primitive cell and will leave no gaps.

General Mathematical Concept

The general mathematical concept embodied in a Wigner-Seitz cell is more commonly called a Voronoi cell. Though the Wigner-Seitz cell in itself is not of paramount importance in direct space, it is extremely important in the reciprocal space. The Wigner-Seitz cell in the reciprocal space is called the Brillouin zone, which contains the information about whether a material will be a conductor, semi-conductor or an insulator.

Xenomorph

A xenomorph (also: allotriomorph) is a mineral that did not develop its otherwise typical external crystal form because of late crystallisation between earlier- formed crystals. Xenomorphs are typical of matrix minerals in rapidly crystallising volcanic lavas and shallow igneous intrusions. It is also typical of the interstitial or cementing minerals formed during the digenesis of sedimentary rocks. The opposite is an idiomorph in which the external form is controlled only by the internal crystal structure.

3

The Minerals

Minerals are natural compounds formed through geological processes. The term 'mineral' encompasses not only the material's chemical composition, but also the mineral's structure.

Minerals range in composition from pure elements and simple salts to very complex silicates with thousands of known forms. The study of minerals is called mineralogy.

Minerals

Definition and Classification

To be classified as a "true" mineral, a substance must be a solid and have a crystalline structure. It must also be a naturally occurring, homogeneous substance with a defined chemical composition. Traditional definitions excluded *organically derived* material. However, the International Mineralogical Association in 1995 adopted a new definition:

> "A mineral is an element or chemical compound that is normally crystalline and that has been formed as a result of geological processes."

The modern classifications include an *organic class* - in both the new Dana and the Srunz classification schemes.

The chemical composition may vary between end members of a mineral system. For example the plagioclase feldspars comprise a continuous series from sodium-rich albite ($NaAlSi_3O_8$) to calcium-rich anorthite ($CaAl_2Si_2O_8$) with four recognised intermediate compositions between. Mineral-like substances that don't strictly meet the definition are sometimes classified as mineraloids. Other natural-occurring substances are non-minerals. Industrial minerals is a market term and refers to commercially valuable mined materials.

A crystal structure is the orderly geometric spatial arrangement of atoms in the internal structure of a mineral. There are 14 basic crystal lattice arrangements of atoms in three dimensions, and these are referred to as the 14 "Bravais lattices". Each of these lattices can be classified into one of the six crystal systems, and all crystal structures currently recognised fit in one Bravais lattice and one crystal system. This crystal structure is based on regular internal atomic or ionic arrangement that is often expressed in the geometric form that the crystal takes. Even when the mineral grains are too small to see or are irregularly shaped, the underlying crystal structure is always periodic, and can be determined by X-ray diffraction.

Chemistry and crystal structure together define a mineral.

In fact, two or more minerals may have the same chemical composition, but differ in crystal structure (these are known as *polymorphs*). For example, pyrite and marcasite are both iron sulphide, but their arrangement of atoms differs. Similarly, some minerals have different chemical compositions, but the same crystal structure: for example, halite (made from sodium and chlorine), galena (made from lead and sulphur) and periclase (made from magnesium and oxygen) all share the same cubic crystal structure.

Crystal structure greatly influences a mineral's physical properties. For example, though diamond and graphite have the same composition (both are pure carbon), graphite is very soft, while diamond is the hardest of all known minerals. This happens because the carbon atoms in graphite are arranged into sheets which can slide easily past each other, while the carbon atoms in diamond form a strong, interlocking three-dimensional network.

There are currently just over 4,000 known minerals, according to the International Mineralogical Association, which is responsible for the approval of and naming of new mineral species found in nature. Of these, perhaps 150 can be called "common," 50 are "occasional," and the rest are "rare" to "extremely rare."

Minerals and Rocks

A mineral is a naturally occurring, inorganic solid with a definite chemical composition and a crystalline structure. A rock is an aggregate of one or more minerals. (A rock may also include organic remains and mineraloids.) Some rocks are predominantly composed of just one mineral. For example, limestone is a sedimentary rock composed almost entirely of the mineral calcite. Other rocks contain many minerals, and the specific minerals in a rock can vary widely. Some minerals, like quartz, mica or feldspar are common, while others have been found in only one or two locations worldwide. The vast majority of the rocks of the Earth's crust consist of quartz, feldspar, mica, chlorite, kaolin, calcite, epidote, olivine, augite, hornblende, magnetite, haematite, limonite and a few other minerals. Over half of the mineral species known are so rare that they have only been found in a handful of samples, and many are known from only one or two small grains.

Commercially valuable minerals and rocks are referred to as industrial minerals. Rocks from which minerals are mined for economic purposes are referred to as ores (the rocks and minerals that remain, after the desired mineral has been separated from the ore, are referred to as tailings).

Mineral Composition of Rocks

A main determining factor in the formation of minerals in a rock mass is the chemical composition of the mass, for a certain mineral can be formed only when the necessary elements are present in the rock. Calcite is most common in limestones, as these consist essentially of carbonate of lime; quartz in sandstones and in certain igneous rocks which contain a high percentage of silica.

Other factors are of equal importance in determining the natural association or paragenesis of rock-making minerals, principally the mode of origin of the rock and the stages through which it has passed in attaining its present condition. Two rock masses may have very much the same bulk composition and yet consist of entirely different assemblages of minerals.

The tendency is always for those compounds to be formed which are stable under the conditions under which the rock mass originated. A granite arises by the consolidation of a molten magma (a fused rock mass) at high temperatures and great pressures and its component minerals are such as are formed in such circumstances.

Exposed to moisture, carbonic acid and other subaerial agents at the ordinary temperatures of the earth's surface, some of these original minerals, such as quartz and white mica are permanent and remain unaffected; others "weather" or decay and are replaced by new combinations.

The felspar passes into kaolin, muscovite and quartz, and if any black mica (biotite) has been present it yields chlorite, apidote, rutile and other substances. These changes are accompanied by disintegration, and the rock falls into a loose, incoherent, earthy mass which may be regarded as a sand or soil. The materials thus formed may be washed away and deposited as a sandstone or grit.

The structure of the original rock is now replaced by a new one; the mineralogical constitution is profoundly altered; but the bulk chemical composition may not be very different. The sedimentary rock may again undergo a metamorphosis.

If penetrated by igneous rocks it may be recrystallised or, if subjected to enormous pressures with heat and movement, such as attend the building of folded mountain chains, it may be converted into a gneiss not very different in mineralogical composition though radically different in structure to the granite which was its original state.

PHYSICAL PROPERTIES OF MINERALS

Classifying minerals can range from simple to very difficult. A mineral can be identified by several physical properties, some of them being sufficient for full identification without equivocation. In other cases, minerals can only be classified by more complex chemical or X-ray diffraction analysis; these methods, however, can be costly and time-consuming.

Physical properties commonly used are:

Crystal Structure and Habit

A mineral may show good crystal habit or form, or it may be massive, granular or compact with only microscopically visible crystals.

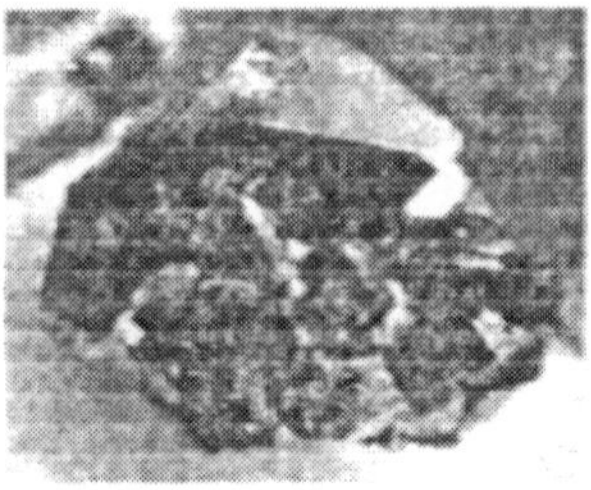

Image— Enargite crystals

In mineralogy and crystallography, a crystal structure is a unique arrangement of atoms in a crystal. A crystal structure is composed of a unit cell, a set of atoms arranged in a particular way, which is periodically repeated in three dimensions on a lattice.

The spacing between unit cells in various directions is called its *lattice parameters.* The symmetry properties of the crystal are embodied in its space group. A crystal's structure and symmetry play a role in determining many of its properties, such as cleavage, electronic band structure, and optical properties.

Unit Cell

The crystal structure of a material is often discussed in terms of its unit cell. The unit cell is a spatial arrangement of atoms which is tiled in three-dimensional space to describe the crystal. The unit cell is given by its lattice parameters, the length of the cell edges and the angles between them, while the positions of the atoms inside the unit cell are described by the set of atomic positions (x_i, y_i, z_i) measured from a lattice point.

For each crystal structure there is a *conventional unit cell,* which is the smallest unit that has the *full symmetry* of the crystal. However, the conventional unit cell is not always the smallest possible choice.

A primitive unit cell of a particular crystal structure is the smallest possible unit cell one can construct such that, when tiled, it completely fills space. This primitive unit cell does not, however, display all the symmetries inherent in the crystal. A Wigner-Seitz cell is a particular kind of primitive cell which has the same symmetry as the lattice.

There are only seven distinct shapes that can pack together into an infinite 3D space lattice in such a way that each lattice point has an identical environment to that around every other lattice point.

Classification of Crystals by Symmetry

The defining property of a crystal is its inherent symmetry, by which we mean that under certain *operations* the crystal remains unchanged. For example, rotating the crystal 180 degrees about a certain axis may result in an atomic configuration which is identical to the original configuration. The crystal is then said to have a twofold rotational symmetry about this axis.

In addition to rotational symmetries like this, a crystal may have symmetries in the form of mirror planes and translational symmetries, and also the so-called *compound symmetries* which are a combination of translation and rotation/mirror symmetries. A full classification of a crystal is achieved when all of these inherent symmetries of the crystal are identified.

Crystal System

The crystal systems are a grouping of crystal structures according to the axial system used to describe their lattice. Each crystal system consists of a set of three axes in a particular geometrical arrangement. There are seven unique crystal systems. The simplest and most symmetric, the cubic (or isometric) system, has the symmetry of a cube, that is, it exhibits four threefold rotational axes oriented at 109.5 degrees (the tetrahedral angle) with respect to each other.

These threefold axes lie along the body diagonals of the cube. This definition of a cubic is correct, although many textbooks incorrectly state that a cube is defined by three mutually perpendicular axes of equal length - if this were true there would be far more than 14 Bravais lattices. The other six systems, in order of decreasing symmetry, are hexagonal, tetragonal, rhombohedral (also known as trigonal), orthorhombic, monoclinic and triclinic. Some crystallographers consider the hexagonal crystal system not to be its own crystal system, but instead a part of the trigonal crystal system. The crystal system and Bravais lattice of a crystal describe the (purely) translational symmetry of the crystal.

Bravais Lattices

When the crystal systems are combined with the various possible lattice centrings, we arrive at the Bravais lattices. They describe the geometric arrangement of the lattice points, and thereby the translational symmetry of the crystal. In three dimensions, there are 14 unique Bravais lattices which are distinct from one another in the translational symmetry they contain. All crystalline materials recognised until now (not including quasicrystals) fit in one of these arrangements. The fourteen, three-dimensional lattices, classified by crystal system, are shown to the right. The Bravais lattices are sometimes referred to as *space lattices*.

The crystal structure consists of the same group of atoms, the *basis*, positioned around each and every lattice point. This group of atoms therefore repeats indefinitely in three dimensions according to the arrangement of one of the 14 Bravais lattices. The characteristic rotation and mirror symmetries of the group of atoms, or unit cell, is described by its crystallographic point group.

Crystal system Lattices:

triclinic

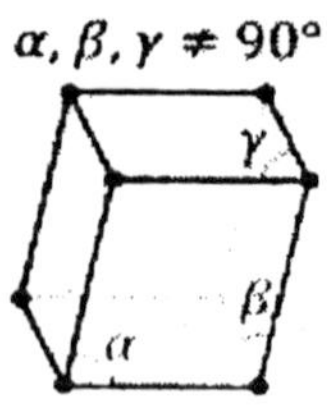

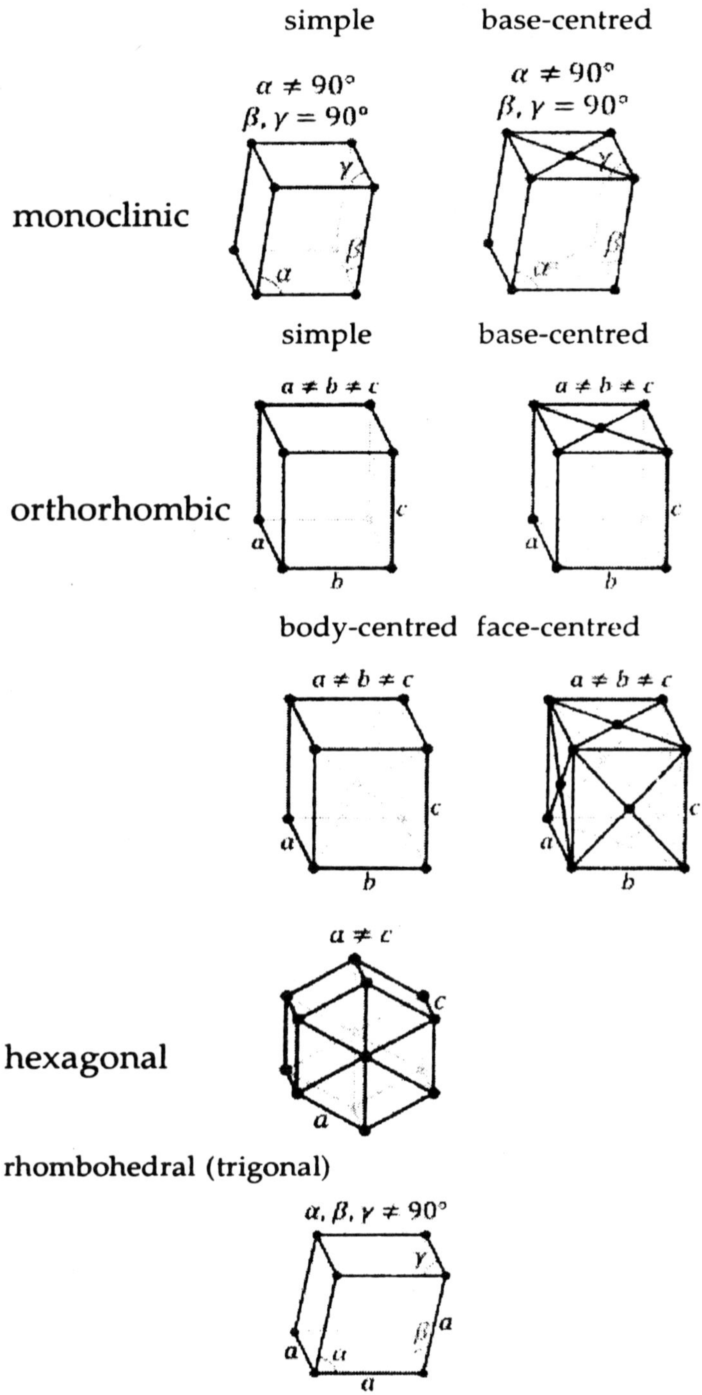
simple
base-centred
α ≠ 90°
β, γ = 90°
α ≠ 90°
β, γ = 90°
monoclinic
simple
base-centred
a ≠ b ≠ c
a ≠ b ≠ c
orthorhombic
body-centred
face-centred
a ≠ b ≠ c
a ≠ b ≠ c
a ≠ c
hexagonal
rhombohedral (trigonal)
α, β, γ ≠ 90°

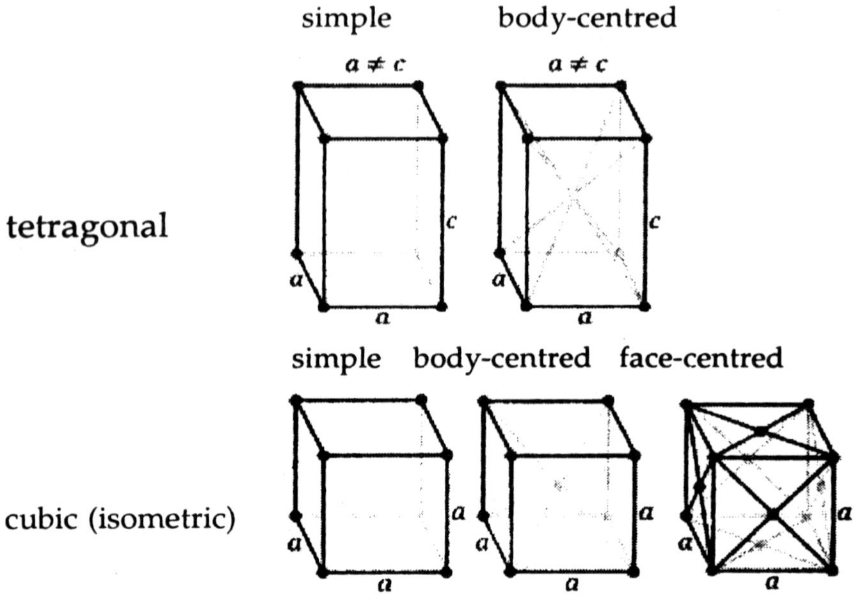

Point and Space Groups

The crystallographic point group or *crystal class* is the set of non-translational symmetry operations that leave the appearance of the crystal structure unchanged.

These symmetry operations can include *mirror planes,* which reflect the structure across a central plane, *rotation axes,* which rotate the structure a specified number of degrees, and a *centre of symmetry* or *inversion point* which inverts the structure through a central point. There are 32 possible crystal classes. Each one can be classified into one of the seven crystal systems.

The space group of the crystal structure is composed of the translational symmetry operations in addition to the operations of the point group.

These include pure *translations* which move a point along a vector, *screw axes,* which rotate a point around an axis while translating parallel to the axis, and *glide planes,* which reflect a point through a plane while translating it parallel to the plane. There are 230 distinct space groups.

Defects in Crystals

Real crystals feature defects or irregularities in the ideal arrangements described above and it is these defects that critically determine many of the electrical and mechanical properties of real materials. In particular dislocations in the crystal lattice allow shear at much lower stress than that needed for a perfect crystal structure.

Crystal Symmetry and Physical Properties

Twenty of the 32 crystal classes are so-called piezoelectric, and crystals belonging to one of these classes (point groups) display piezoelectricity. All 20 piezoelectric classes lack a centre of symmetry. Any material develops a dielectric polarisation when an electric field is applied, but a substance which has such a natural charge separation even in the absence of a field is called a polar material.

Whether or not a material is polar is determined solely by its crystal structure. Only 10 of the 32 point groups are polar. All polar crystals are pyroelectric, so the 10 polar crystal classes are sometimes referred to as the pyroelectric classes.

There are a few crystal structures, notably the perovskite structure, which exhibit ferroelectric behaviour. This is analogous to ferromagnetism, in that, in the absence of an electric field during production, the ferroelectric crystal does not exhibit a polarisation.

Upon the application of an electric field of sufficient magnitude, the crystal becomes permanently polarised. This polarisation can be reversed by a sufficiently large counter-charge, in the same way that a ferromagnet can be reversed. However, it is important to note that, although they are called ferroelectrics, the effect is due to the crystal structure, not the presence of a ferrous metal.

Incommensurate crystals have period-varying translational symmetry. The period between nodes of symmetry is constant in most crystals. The distance between nodes in an incommensurate crystal is dependent on the number of nodes between it and the base node.

Hardness

The physical hardness of a mineral is usually measured according to the Mohs scale. This scale is relative and goes from 1 to 10. Minerals with a given Mohs hardness can scratch the surface of any mineral that has a lower hardness than itself.

Mohs Scale of Mineral Hardness

The Mohs scale of mineral hardness characterises the scratch resistance of various minerals through the ability of a harder material to scratch a softer material. It was created in 1812 by the German mineralogist Friedrich Mohs and is one of several definitions of hardness in materials science.

Mohs based the scale on ten minerals that are all readily available. As the hardest known naturally occurring substance, diamond is at the top of the scale. The hardness of a material is measured against the scale by finding the hardest material that the given material can scratch, and/or the softest material that can scratch the given material. For example, if some material is scratched by apatite but not by fluorite, its hardness on the Mohs scale is 4.5.

The Mohs scale is a purely ordinal scale. For example, corundum (9) is twice as hard as topaz (8), but diamond (10) almost four times as hard as corundum. The table below shows comparison with absolute hardness measured by a sclerometer.

Hardness	*Mineral*	*Absolute Hardness*
1	Talc ($Mg_3Si_4O_{10}(OH)_2$)	1
2	Gypsum ($CaSO_4 \cdot 2H_2O$)	2
3	Calcite ($CaCO_3$)	9
4	Fluorite (CaF_2)	21
5	Apatite ($Ca_5(PO_4)_3(OH^-,Cl^-,F^-)$)	48
6	Orthoclase Feldspar ($KAlSi_3O_8$)	72

contd........

Hardness	Mineral	Absolute Hardness
7	Quartz (SiO_2)	100
8	Topaz ($Al_2SiO_4(OH^-,F^-)_2$)	200
9	Corundum (Al_2O_3)	400
10	Diamond (C)	1500

On the Mohs scale, fingernail has hardness 2.5; copper penny, about 3.5; a knife blade, 5.5; window glass, 6.5; steel file, 6.5. Using these ordinary materials of known hardness can be a simple way to approximate the position of a mineral on the scale. Some mnemonics traditionally taught to geology students to remember this table are "The Girls Can Flirt And Other Queer Things Can Do" or "To Get Candy From Aunt Fanny, Quit Teasing Cousin Danny". An alternative table is shown below which has been modified to incorporate additional substances that may fall in between two levels.

Hardness	Substance or Mineral
1	Talc
2	Gypsum
2.5 to 3	Pure Gold, Silver
3	Calcite, Copper penny
4	Fluorite
4 to 4.5	Platinum
4 to 5	Iron
5	Apatite
6	Orthoclase
6.5	Iron pyrite
6 to 7	Glass, Vitreous pure silica
7	Quartz
7 to 7.5	Garnet
7 to 8	Hardened steel
8	Topaz
9	Corundum
10	Diamond
11.1	Aggregated diamond nanorods

- *Lustre* indicates the way a mineral's surface interacts with light and can range from dull to glassy (vitreous).
- Metallic -high reflectivity like metal: galena and pyrite
- Sub-metallic -slightly less than metallic reflectivity: magnetite
- Non-metallic lustres:
- *Adamantine* - brilliant, the lustre of diamond also cerussite and anglesite
- *Vitreous* -the lustre of a broken glass: quartz
- *Pearly* - iridescent and pearl-like: talc and apophyllite
- *Resinous* - the lustre of resin: sphalerite and sulphur
- *Silky* - a soft light shown by fibrous materials: gypsum and chrysotile
- Dull/earthy -shown by finely crystallised minerals: the kidney ore variety of haematite
- *Colour* indicates the appearance of the mineral in reflected light or transmitted light for translucent minerals (i.e. what it looks like to the naked eye).
- *Irridescence* - the *play of colours* due to surface or internal interference. Labradorite exhibits internal irridescence whereas haematite and sphalerite often show the surface effect.
- *Streak* refers to the colour of the powder a mineral leaves after rubbing it on an unglazed porcelain *streak plate*. Note that this is not always the same colour as the original mineral.
- *Cleavage* describes the way a mineral may split apart along various planes. In thin section, cleavage is visible as thin parallel lines across a mineral.
- *Fracture* describes how a mineral breaks when broken contrary to its natural cleavage planes.
- *Chonchoidal fracture* is a smooth curved fracture with concentric ridges of the type shown by glass.
- *Hackley* is jagged fracture with sharp edges.

STREAK

The streak (also called powder colour) of a mineral is the colour of the powder produced when it is dragged across an unweathered surface. Unlike the apparent colour of a mineral, which for most minerals can vary considerably, the trail of finely ground powder generally has a more consistent characteristic colour, and is thus an important diagnostic tool in mineral identification. If no streak seems to be made, the mineral's streak is said to be white or colourless. Streak is particularly important as a diagnostic for opaque and coloured materials. It is less useful for silicates, most of which have a white streak and are too hard to powder easily.

The apparent colour can vary widely because of trace impurities or a disturbed macroscopic crystal structure. Small amounts of an impurity that strongly absorbs a particular wavelength can radically change the wavelengths of light that are reflected by the specimen, and thus change the apparent colour. However, when the specimen is dragged to produce a streak, it is broken into randomly oriented microscopic crystals, and small impurities do not greatly affect the absorption of light.

The surface across which the mineral is dragged is called a "streak plate," and is generally made of unglazed porcelain tile. In the absence of a streak plate, the underside of a porcelain bowl or vase, the surface of an electrical fuse, or the back of a glazed tile will work. Sometimes a streak is more easily or accurately described by comparing it with the "streak" made by another streak plate.

Because the trail left behind results from the mineral being crushed into powder, a streak can only be made of minerals softer than the streak plate, around 7 on the Mohs scale of mineral hardness. In this case, the colour of the powder can be determined by filing or crushing with a hammer a small sample, which is then usually rubbed on a streak plate. Most minerals that are harder have an unhelpful white streak.

Some minerals leave a streak similar to their natural colour, such as cinnabar and azurite. Other minerals leave surprising colours, such as fluorite, which always has a white streak, although it can appear in purple, blue, yellow, or green crystals. Haematite, which is black in appearance, leaves a red streak which accounts for its name, which comes from the Greek word "haima," meaning "blood." Galena, which can be similar in appearance to haematite, is easily distinguished by its gray streak.

CLEAVAGE (CRYSTAL)

Cleavage, in mineralogy, is the tendency of crystalline materials to split along definite planes, creating smooth surfaces, of which there are several named types:

- *Basal Cleavage:* cleavage parallel to the base of a crystal, or to the plane of the lateral axes. This occurs quite easily in graphite, making the material feel slippery.
- *Cubic Cleavage:* cleavage parallel to the faces of a cube. This is the source of the cubic shape seen in crystals of ground table salt (sodium chloride).
- *Diagonal Cleavage:* cleavage parallel to a diagonal plane.
- *Lateral Cleavage:* cleavage parallel to the lateral planes.
- *Octahedral, Dodecahedral, or Rhombohedral cleavage:* cleavage parallel to the faces of an octahedron, dodecahedron, or rhombohedron (respectively). Octahedral cleavage is seen in common semi-conductors.

Prismatic Cleavage, Cleavage Parallel to a Vertical Prism

This is of technical importance in the electronics industry and in the cutting of gemstones. While precious stones are generally cleaved by impact, man-made single crystals of semi-conductor materials are generally sold as thin wafers which are much easier to cleave. Simply pressing a silicon wafer against a soft surface and scratching its edge with a diamond scribe is usually enough to cause cleavage; however, when dicing a wafer to form chips, a procedure of scoring and breaking is often followed for greater control. Elemental semi-conductors (Si, Ge, and diamond) are diamond cubic, a space group for which octahedral cleavage is observed. This means that some orientations of wafer allow near-perfect rectangles to be cleaved. Most other commercial semi-conductors (GaAs, InSb, etc.) can be made in the related zinc blende structure, with similar cleavage planes.

SPECIFIC GRAVITY

The term specific gravity is similar, except that the reference material is water. A relative density can help quantify the buoyancy between two materials, or determine the density of one "unknown" material using the "known" density of another material. Mathematically, relative density is expressed as:

$$G = \frac{\text{Pobject}}{\text{Pr eference}}$$

where *G* is the relative density, and *P* is the densities of the two materials in the same units (e.g., kg/m^3, g/cm^3).

Relative density is dimensionless, since it is a ratio between two quantities of same unit. If the ratio is greater than 1, the *object* will be heavier than the same volume of the *reference*. If it is less than 1, it will be lighter than the *reference*. It is important to specify the reference material when reporting a relative density, but when the reference material is not specified it is usually understood to be water at 3.98 ° C.

Difference between Specific Gravity and Relative Density

Specific gravity is a special case of relative density. While specific gravity has a reference density of water, relative density can have any reference density that is used. It is best to specify the reference material when using relative density, using subscripts:

$$RD_{\text{Object/Reference}}$$

Which states "the relative density of the object with respect to the reference material".

While relative density will not change as long as consistent units are maintained, the relative density is relative to its reference. The relative density of an object relative to mercury, is different than the relative density with respect to water (specific gravity).

Taking the relative density with respect to alcohol, the water has a specific gravity of 1.2, and the iron has a specific gravity of 10. Taking the relative density relative to water, the numbers would be.78, 1, and 7.9. With respect to iron, the numbers are .1, .12, and 1. These are all correct relative densities, however it is with different reference points. Note that again, units are not needed and hence these numbers are correct no matter what unit system is used.

Practical Uses of the Unitless Properties of Relative Density

Maintaining a consistent set of units is important. The units do not need to even be in terms of "amount of substance" per "volume" directly. For example, suppose someone wanted to find out the specific gravity of a rock. They could see that the rock deflected a certain spring by 3 inches.

Then they put on a reference substance, which deflected the spring by 5 inches. The first rock made the water in a certain graduated cylinder rise by 20 mm, and the reference made it rise by 34 mm. The relative density between these two objects can easily be determined without having to figure out several constants which would be needed to determine the density directly (like the spring constant or the cross-sectional area of the cylinder).

Since they know that the density of each object would be:

$$\rho\text{object} = \frac{\text{Mass}}{\text{Volume}} = \frac{\text{Deflection} * \text{Spring Cons tan t} * \text{Gravity}}{\text{Dsplacement}_{\text{water Line}} * \text{Area}_{\text{Cy ln der}}}$$

Then the relative density is:

$$\text{RD} = \frac{\rho\text{object}}{\rho H_2O} = \frac{\dfrac{\text{Deflection}_{\text{obj}} * \text{springCons tan t} * \text{Gravity}}{\text{Displacement}_{\text{obj}} * \text{Area Cylinder}}}{\dfrac{\text{Deflection Re f.} * \text{Spring Cons tan ts} * \text{Gravity}}{\text{Displacements Re f.} * \text{Cylinder}}}$$

$$RD = \frac{\frac{\text{Deflection}_{obj.}}{\text{Displacement}_{obj}}}{\frac{\text{Deflection Ref.}}{\text{Displacements Ref.}}} = \frac{\frac{3\ in}{20\ mm}}{\frac{5\ in}{34\ mm}} = \frac{3in * 34\,mm}{5in * 20\ mm} = 1.02$$

Now if they knew the actual density of the reference substance, then they can find the actual density of the test object. In this example, the density of the reference substance is 300000 N/inch³ (which would be absurd, but shows the power of the RD system). Then the density of the test substance is 300000 N/in³ * 1.02 = 306000 N/in³.

Temperature Dependence

Changes in temperature affect the densities of materials, and the relative densities between two materials need to be corrected at certain temperatures. It is best to record the temperatures of the two materials, expressed here:

relative density: $8.15^{20°C}_{4°C}$ or specific gravity: 2.432^{15}_{0}

where the superscript indicate the temperature at which the density of the material is measured, and the subscripts indicate the temperature of the water to which it is compared.

Uses of Relative Density

Relative density is often used by geologists and mineralogists to help determine the mineral content of a rock or other sample. Gemologists use it as an aid in the identification of gemstones. The reason that relative density is measured in terms of the density of water is because that is the easiest way to measure it in the field.

Basically, density is defined as the mass of a sample divided by its volume. With an irregularly shaped rock, the volume can be very difficult to accurately measure. One way is to put it in a water-filled graduated cylinder and see how much water it displaces. Relative density is more easily and perhaps more accurately measured without measuring volume. Simply suspend the sample from a spring scale and weigh it under water. The following formula is used to determine the specific gravity:

$$G = \frac{W}{W - F}$$

where:

G is the relative density,

W is the weight of the sample (measured in pounds-force, newtons, or some other unit of force),

F is the force, measured in the same units, while the sample was submerged.

Note that with this technique it is difficult to measure relative densities less than one, because in order to do so, the sign of F must change, requiring the measurement of the downward force needed to keep the sample underwater.

Another practical method uses three measurements. The mineral sample is weighed dry. Then a container filled to the brim with water is weighed, and weighed again with the sample immersed, after the displaced water has overflowed and been removed. Subtracting the last reading from the sum of the first two readings gives the weight of the displaced water. The relative density result is the dry sample weight divided by that of the displaced water.

This method works with scales that can't easily accommodate a suspended sample, and also allows for measurement of samples that are less dense than water. Surface tension of the water may keep a significant amount of water from overflowing, which is especially problematic for small objects being immersed. A workaround would be to use a water container with as small a mouth as possible.

Relative density is not important physically as much as it is in reporting data to the masses. It is difficult to explain the high density of iridium as simply being 22650 kg/m^3. To the masses, this number is just a large number. But if the facts instead detailed that the density of iridium is nearly twice that of lead, people can understand how heavy iridium really is. As a reporter, the relative density tells more information to the general public then the actual number does.

GEMOLOGY, MINERALOGY, GEOLOGY AND FORENSICS

Gemstones, minerals, fibres and many other materials which may be encountered in forensics or with a relationship to various collectibles may have a distinctive fluorescence or may fluoresce differently under short-wave ultraviolet, long-wave ultra violet, or X-rays.

Many types of calcite and amber will fluoresce under shortwave UV. Rubies, emeralds, and the Hope Diamond exhibit red fluorescence under short-wave UV light; diamonds also emit light under X ray radiation. Fluorescence can also be used to help recognise chirality in minerals.

Crude oil (Petroleum) fluoresces in a range of colours, from dull brown for heavy oils and tars through to bright yellowish and bluish white for very light oils and condensates. This phenomenon is used in oil exploration drilling to identify very small amounts of oil in drill cuttings and core sample.

PIEZOELECTRICITY

Piezoelectricity is the ability of crystals and certain ceramic materials to generate a voltage in response to applied mechanical stress. Piezoelectricity was discovered by Pierre Curie and the word is derived from the Greek piezein, which means to squeeze or press. The piezoelectric effect is reversible in that piezoelectric crystals, when subjected to an externally applied voltage, can change shape by a small amount. (For instance, the deformation is about 0.1 per cent of the original dimension in PZT.)

The effect finds useful applications such as the production and detection of sound, generation of high voltages, electronic frequency generation, microbalance, and ultra fine focusing of optical assemblies.

A related property known as pyroelectricity, the ability of certain mineral crystals to generate electrical charge when heated, was known of as early as the 19th century, and was named by David Brewster in 1824.

In 1880, the brothers Pierre Curie and Jacques Curie predicted and demonstrated piezoelectricity using tinfoil, glue, wire, magnets, and a jeweller's saw. They showed that crystals of tourmaline, quartz, topaz, cane sugar, and Rochelle salt (sodium potassium tartrate tetrahydrate) generate electrical polarisation from mechanical stress. Quartz and Rochelle salt exhibited the most piezoelectricity.

Converse piezoelectricity was mathematically deduced from fundamental thermodynamic principles by Lippmann in 1881. The Curies immediately confirmed the existence of the "converse effect," and went on to obtain quantitative proof of the complete reversibility of electro-elasto-mechanical deformations in piezoelectric crystals.

World War I and Piezoelectric Devices

The second application for piezoelectric devices was sonar, first developed during World War I. In France in 1917, Paul Langevin (whose development now bears his name) and his co-workers developed an ultrasonic submarine detector. The detector consisted of a transducer, made of thin quartz crystals carefully glued between two steel plates, and a hydrophone to detect the returned echo. By emitting a high-frequency chirp from the transducer, and measuring the amount of time it takes to hear an echo from the sound waves bouncing off an object, one can calculate the distance to that object.

The use of piezoelectricity in sonar, and the success of that project, created intense development interest in piezoelectric devices. Over the next few decades, new piezoelectric materials and new applications for those materials were explored and developed.

Development of piezoelectric devices and materials in the United States was kept within the companies doing the development, mostly due to the wartime beginnings of the field, and in the interests of securing profitable patents. New materials were the first to be developed — quartz crystals were the first commercially exploited piezoelectric material, but scientists searched for higher-performance materials.

Piezoelectric devices found homes in many fields. Ceramic phonograph cartridges simplified player design, were cheap and accurate, and made record players cheaper to maintain and easier to build. Ceramic electret microphones could be made small and sensitive. The development of the ultrasonic transducer allowed for easy measurement of viscosity and elasticity in fluids and solids, resulting in huge advances in materials research.

Ultrasonic time-domain reflectometers (which send an ultrasonic pulse through a material and measure reflections from discontinuities) could find flaws inside cast metal and stone objects, improving structural safety. However, despite the advances in materials and the maturation of manufacturing processes, the United States market had not grown as quickly. Without many new applications, the growth of the United States' piezoelectric industry suffered.

In contrast, Japanese manufacturers shared their information, quickly overcoming technical and manufacturing challenges and creating new markets. Japanese efforts in materials research created piezoceramic materials competitive to the US materials, but free of expensive patent restrictions. Major Japanese piezoelectric developments include new designs of piezoceramic filters, used in radios and televisions, piezo buzzers and audio transducers that could be connected directly into electronic circuits, and the piezoelectric igniter which generates sparks for small engine ignition systems (and gas-grill lighters) by compressing a ceramic disc. Ultrasonic transducers that could transmit sound waves through air had existed for quite some time, but first saw major commercial use in early television remote controls. These transducers now are mounted on several car models as an echolocation device, helping the driver determine the distance from the rear of the car to any objects that may be in its path.

Materials

In addition to crystals of tourmaline, quartz, topaz, cane sugar, and Rochelle salt, many other materials exhibit the effect, including quartz analogue crystals like berlinite ($AlPO_4$) and gallium orthophosphate ($GaPO_4$), ceramics with perovskite or tungsten-bronze structures ($BaTiO_3$, $SrTiO_3$, $Pb(ZrTi)O_3$, $KNbO_3$, $LiNbO_3$, $LiTaO_3$, $BiFeO_3$, Na_xWO_3, $Ba_2NaNb_5O_5$, $Pb_2KNb_5O_{15}$).

Polymer materials like rubber, wool, hair, wood fibre, and silk often behave as electrets. Although this phenomenon is often confused with piezoelectricity, the two phenomena are distinct. The orientation of polarisation in a piezoelectric is limited by the symmetry, whereas the polarisation direction in an electret is not. The polymer polyvinylidene fluoride, PVDF, exhibits piezoelectricity several times larger than quartz. Bone exhibits some piezoelectric properties, due to the apatite crystals: it has been hypothesised that this is part of the mechanism of bone remodelling in response to stress, as the electric fields on the apatite crystals stimulate further bone growth.

Applications

Piezoelectric crystals are now used in numerous ways:

High-voltage Sources

Direct piezoelectricity of some substances like quartz, as mentioned above, can generate potential differences of thousands of volts.

Diagnostic Ultrasound Medical Ultrasonography

Probably the best-known application is the electric cigarette lighter:

> Pressing the button causes a spring-loaded hammer to hit a piezoelectric crystal, and the high voltage produced ignites the gas as the current jumps over a small spark gap. The portable sparkers used to light gas grills or stoves work the same way.

A similar idea is being researched by DARPA in the United States in a project called *Energy Harvesting*, which includes an attempt to power battlefield equipment by piezoelectric generators embedded in soldiers' boots.

A piezoelectric transformer is a type of AC voltage multiplier. Unlike a conventional transformer, which uses magnetic coupling between input and output, the piezoelectric transformer uses acoustic coupling. An input voltage is applied across a short length of a bar of piezoceramic material such as PZT, creating an alternating stress in the bar by the inverse piezoelectric effect and causing the whole bar to vibrate. The vibration frequency is chosen to be the resonant frequency of the block, typically in the 100 kilohertz to 1 megahertz range. A higher output voltage is then generated across another section of the bar by the piezoelectric effect. Step-up ratios of more than 1000:1 have been demonstrated. An extra feature of this transformer is that, by operating it above its resonant frequency, it can be made to appear as an inductive load, which is useful in circuits that require a controlled soft start.

Sensors

The principle of operation of a piezoelectric sensor is that a physical dimension, transformed into a force, acts on two opposing faces of the sensing element. Depending on the design of a sensor, different "modes" to load the piezoelecric element can be used: longitudinal, transversal and shear.

To detect sound, e.g. piezoelectric microphones (sound waves bend the piezoelectric material, creating a changing voltage) and piezoelectric pickups for electrically amplified guitars. A piezo sensor attached to the body of an instrument is known as a contact microphone. Piezoelectric elements are also used in the generation of sonar waves. Piezoelectric microbalances are used as very sensitive chemical and biological sensors.

Piezoelectric transducers are used in electronic drum pads to detect the impact of the drummer's sticks.

Automotive engine management systems use a piezoelectric transducer to detect detonation, by sampling the vibrations of the engine block.

Actuators

As very high voltages correspond to only tiny changes in the width of the crystal, this width can be changed with better-than-micrometer precision, making piezo crystals the most important tool for positioning objects with extreme accuracy—thus their use in actuators.

Loudspeakers

Voltages are converted to mechanical movement of a piezoelectric polymer film.

Piezoelectric Motors

Piezoelectric elements apply a directional force to an axle, causing it to rotate. Due to the extremely small distances involved, the piezo motor is viewed as a high-precision replacement for the stepper motor.

Piezoelectric elements can be used in laser mirror alignment, where their ability to move a large mass (the mirror mount) over microscopic distances is exploited to electronically align some laser mirrors. By precisely controlling the distance between mirrors, the laser electronics can accurately maintain optical conditions inside the laser cavity to optimise the beam output.

A related application is the acousto-optic modulator, a device that vibrates a mirror to give the light reflected off it a Doppler shift. This is useful for fine-tuning a laser's frequency.

Atomic force microscopes and scanning tunnelling microscopes employ converse piezoelectricity to keep the sensing needle close to the probe.

Inkjet Printers

On many inkjet printers; piezoelectric crystals are used to control the flow of ink from the cartridge to the paper.

Diesel Engines

High-performance common rail diesel engines use piezoelectric fuel injectors, first developed by Siemens AG, instead of the more common solenoid valve devices.

Frequency Standard

The piezoelectrical properties of quartz are useful as standard of frequency.

Quartz clocks employ a tuning fork made from quartz that uses a combination of both direct and converse piezoelectricity to generate a regularly timed series of electrical pulses that is used to mark time. The quartz crystal (like any elastic material) has a precisely defined natural frequency (caused by its shape and size) at which it prefers to oscillate, and this is used to stabilise the frequency of a periodic voltage applied to the crystal. The same principle is critical in all radio transmitters and receivers, and in computers where it creates a clock pulse. Both of these usually use a frequency multiplier to reach the megahertz and gigahertz ranges.

Piezoelectric Motors

Types of piezoelectric motor include the well-known travelling-wave motor used for auto-focus in reflex cameras, inchworm motors for linear motion, and rectangular four-quadrant motors with high power density (2.5 watt/cm^3) and speed ranging from 10 nm/s to 800 mm/s. All these motors work on the same principle. Driven by dual orthogonal vibration modes with a phase shift of 90°, the contact point between two surfaces vibrates in an elliptical path, producing a frictional force between the surfaces. Usually, one surface is fixed causing the other to move. In most piezoelectric motors the piezoelectric crystal is excited by a sine wave signal at the resonant frequency of the motor. Using the resonance effect, a much lower voltage can be used to produce a high vibration amplitude.

Ultrasonic Transducers

Piezoelectric materials are used as ultrasonic transducers for imaging applications (e.g. medical imaging, industrial non-destructive testing, or NDT) and high power applications (e.g. medical treatment, sonochemistry and industrial processing).

For imaging applications, the transducer can act as both a sensor and an actuator. Ultrasonic transducers can inject ultrasound waves into the body, receive the returned wave, and convert it to an electrical signal (a voltage). Most medical ultrasound transducers are piezoelectric.

When several elements are stacked upon each other and to the end a blade is placed we can drive the crystal and with the displacement achieved we will have an ultrasonic cutter. This kind of tool is useful when cutting plastic materials or similars as soft material doesn't get affected by the ultrasonic vibration; in fact the cut happens to a local melting of the material due to friction and no application of heat.

Reduction of Vibrations

The TU Darmstadt in Germany researches ways to reduce and stop vibrations by attaching piezo elements. When the material is bent by a vibration in one direction, the system observes the bend and sends electric power to the piezo element to bend in the other direction.

Such an experiment was shown at the Material Vision Fair in Frankfurt in November 2005. Several panels were hit with a rubber mallet, and the panel with the piezo element immediately stopped swinging. The research team sees future applications in cars and houses to reduce noise.

CRYSTAL CLASSES

Of the thirty-two crystal classes, twenty-one are non-centrosymmetric (not having a centre of symmetry), and of these, twenty exhibit direct piezoelectricity (the 21st is the cubic class 432). Ten of these are polar (i.e. spontaneously polarise), having a dipole in their unit cell, and exhibit pyroelectricity. If this dipole can be reversed by the application of an electric field, the material is said to be ferroelectric.

- Piezoelectric Crystal Classes: 1, 2, m, 222, mm2, 4, -4, 422, 4mm, -42m, 3, 32, 3m, 6, -6, 622, 6mm, -62m, 23,-43m
- Pyroelectric: 1, 2, m, mm2, 4, 4mm, 3, 3m, 6, 6mm

In a piezoelectric crystal, the positive and negative electrical charges are separated, but symmetrically distributed, so that the crystal overall is electrically neutral. Each of these sites forms an electric dipole and dipoles near each other tend to be aligned in regions called Weiss domains.

The domains are usually randomly oriented, but can be aligned during polling (not the same as magnetic polling), a process by which a strong electric field is applied across the material, usually at elevated temperatures.

When a mechanical stress is applied, this symmetry is disturbed, and the charge asymmetry generates a voltage across the material. For example, a 1 cm cube of quartz with 500 lbf (2 kN) of correctly applied force upon it, can produce a voltage of 12,500 V.

Piezoelectric materials also show the opposite effect, called inverse piezoelectricity, where the application of an electrical field creates mechanical deformation in the crystal.

MATHEMATICAL DESCRIPTION

Piezoelectricity is the combined effect of the electrical behaviour of the material:

$$D = \in E$$

Where D is the electric displacement, $\in$ is permittivity and E is electric field strength, and Hooke's Law:

$$S = sT$$

Where S is strain, s is compliance and T is stress.

These may be combined into so-called *coupled equations*, of which the strain-charge form is:

$$\{S\} = \left[s^E\right]\{T\} + [d_t]\{E\}$$

$$\{D\} = [d]\{T\} + \left[\in^T\right]\{E\}$$

where d represents the piezoelectric constants, and the superscript E indicates a zero, or constant, electric field; the superscript T indicates a zero, or constant, stress field; and the subscript t stands for transposition of a matrix.

The strain-charge for a material of the 6mm crystal class (such as a polled piezoelectric ceramic, e.g. PZT) may also be written as:

$$\begin{bmatrix} S_1 \\ S_2 \\ S_3 \\ S_4 \\ S_5 \\ S_6 \end{bmatrix} = \begin{bmatrix} s_{11}^E & s_{12}^E & s_{13}^E & 0 & 0 & 0 \\ s_{12}^E & s_{11}^E & s_{13}^E & 0 & 0 & 0 \\ s_{13}^E & s_{13}^E & s_{33}^E & 0 & 0 & 0 \\ 0 & 0 & 0 & s_{44}^E & 0 & 0 \\ 0 & 0 & 0 & 0 & s_{44}^E & 0 \\ 0 & 0 & 0 & 0 & 0 & s_{66}^E = 2\left(s_{11}^E - s_{12}^E\right) \end{bmatrix} \begin{bmatrix} T_1 \\ T_2 \\ T_3 \\ T_4 \\ T_5 \\ T_6 \end{bmatrix} + \begin{bmatrix} 0 & 0 & d_{31} \\ 0 & 0 & d_{31} \\ 0 & 0 & d_{33} \\ 0 & d_{15} & 0 \\ d_{15} & 0 & 0 \\ 0 & 0 & 0 \end{bmatrix} \begin{bmatrix} E_1 \\ E_2 \\ E_3 \end{bmatrix}$$

$$\begin{bmatrix} D_1 \\ D_2 \\ D_3 \end{bmatrix} = \begin{bmatrix} 0 & 0 & 0 & 0 & d_{15} & 0 \\ 0 & 0 & 0 & d_{15} & 0 & 0 \\ d_{31} & d_{31} & d_{33} & 0 & 0 & 0 \end{bmatrix} \begin{bmatrix} T_1 \\ T_2 \\ T_3 \\ T_4 \\ T_5 \\ T_6 \end{bmatrix} + \begin{bmatrix} \epsilon_{11} & 0 & 0 \\ 0 & \epsilon_{11} & 0 \\ 0 & 0 & \epsilon_{33} \end{bmatrix} \begin{bmatrix} E_1 \\ E_2 \\ E_3 \end{bmatrix}$$

The bending forces generated by converse piezoelectricity are extremely high, of the order of tens of millions of pounds (tens of meganewtons), and usually cannot be constrained. The only reason the force is usually not noticed is because it causes a displacement of the order of a few nanometres (billionth of an inch).

Although the above equations are the most used form in literature, some comments about the notation are necessary. Generally *D* and *E* are vectors, that is, Cartesian tensor of rank-1; and permittivity is Cartesian tensor of rank-2. Strain and stress are, in principle, also rank-2 tensors.

But conventionally, because strain and stress are all symmetric tensors, the substript of strain and stress can be relabelled in the following fashion: $11 \rightarrow 1; 22 \rightarrow ; 33 \rightarrow 3; 23 \rightarrow 4; 13 \rightarrow 5; 12 \rightarrow ; 12 \rightarrow 6$. (Different convention may be used by different authors in literature. Say, some use $12 \rightarrow 4$; $23 \rightarrow 5$; $31 \rightarrow 6$ instead.) That is why *S* and *T* appear to have the "vector form" of 6 components. Consequently, *s* appears to be a 6 by 6 matrix instead of rank-4 tensor. Such a relabelled notation is often called Voigt notation.

Chemical Properties of Minerals

Minerals may be classified according to chemical composition. They are here categorised by anion group.

Silicate Class

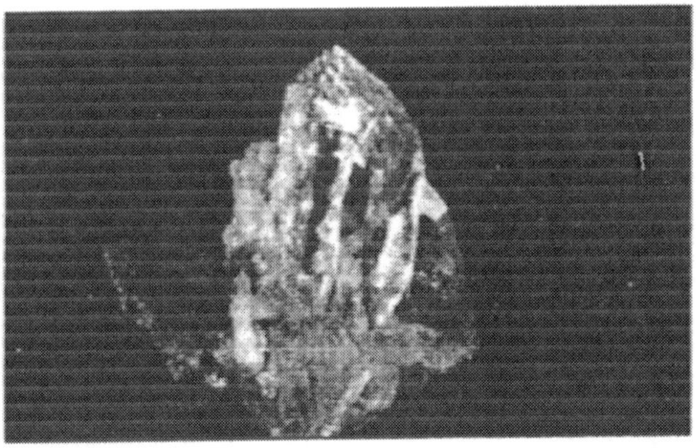

Quartz Crystal

The largest group of minerals by far are the silicates (most rocks are >95 per cent silicates), which are composed largely of silicon and oxygen, with the addition of ions such as aluminium, magnesium, iron, and calcium. Some important rock-forming silicates include the feldspars, quartz, olivines, pyroxenes, amphiboles, garnets, and micas.

Silicate Minerals

The silicate minerals make up the largest and most important class of rock-forming minerals. They are classified based on the structure of their silicate ion group.

Nesosilicates or Isosilicates

Nesosilicates (or *orthosilicates*) have $[SiO_4]^{4-}$ tetrahedra that are isolated and connected by interstitial cations.

Kyanite Crystals

Phenacite group

- Phenacite - Be_2SiO_4
- Willemite - Zn_2SiO_4

Olivine group

- Forsterite - Mg_2SiO_4
- Fayalite - Fe_2SiO_4

Garnet group

- Pyrope - $Mg_3Al_2(SiO_4)_3$
- Almandine - $Fe_3Al_2(SiO_4)_3$
- Spessartine - $Mn_3Al_2(SiO_4)_3$
- Grossular - $Ca_3Al_2(SiO_4)_3$
- Andradite - $Ca_3Fe_2(SiO_4)_3$
- Uvarovite - $Ca_3Cr_2(SiO_4)_3$
- Hydrogrossular - $Ca_3Al_2Si_2O_8(SiO_4)_{1-m}(OH)_{4m}$

Zircon group

- Zircon - Zr_2SiO_4
- Thorite - $(Th,U)_2SiO_4$
- Al_2SiO_5 group

- Andalusite - Al_2SiO_5
- Kyanite - Al_2SiO_5
- Sillimanite - Al_2SiO_5
- Dumortierite - $Al_{6.5-7}BO_3(SiO_4)_3(O,OH)_3$
- Topaz - $Al_2SiO_4(F,OH)_2$
- Staurolite - $Fe_2Al_9(SiO_4)_4(O,OH)_2$

Humite group - $(Mg,Fe)_7(SiO_4)_3(F,OH)_2$

- Norbergite - $Mg_3(SiO_4)(F,OH)_2$
- Chondrodite - $Mg_5(SiO_4)(F,OH)_2$
- Humite - $Mg_7(SiO_4)(F,OH)_2$
- Clinohumite - $Mg_9(SiO_4)(F,OH)_2$
- Datolite - $CaBSiO_4(OH)$
- Titanite - $CaTiSiO_5$
- Chloritoid - $(Fe,Mg,Mn)_2Al_4Si_2O_{10}(OH)_4$

SOROSILICATES

Sorosilicates have isolated double tetrahedra groups with $(Si_2O_7)^{6-}$ or a ratio of 2:7.

- Hemimorphite (calamine) - $Zn_4(Si_2O_7)(OH)_2 \cdot H_2O$
- Lawsonite - $CaAl_2(Si_2O_7)(OH)_2 \cdot H_2O$
- Ilvaite - $CaFe^{2+}{}_2Fe^{3+}O(Si_2O_7)(OH)$
- Epidote group (has both $(SiO_4)^{4-}$ and $(Si_2O_7)^{6-}$ groups)
- Epidote - $Ca_2(Al,Fe)_3O(SiO_4)(Si_2O_7)(OH)$
- Zoisite - $Ca_2Al_3O(SiO_4)(Si_2O_7)(OH)$
- Clinozoisite - $Ca_2Al_3O(SiO_4)(Si_2O_7)(OH)$
- Tanzanite - $Ca_2Al_3O(SiO_4)(Si_2O_7)(OH)$
- Allanite - $Ca(Ce,La,Y,Ca)Al_2(Fe^{2+},Fe^{3+})O(SiO_4)(Si_2O_7)(OH)$
- Vesuvianite(idocrase)-$Ca_{10}(Mg,Fe)_2\ Al_4\ (SiO_4)_5\ (Si_2O_7)_2(OH)_4$

Cyclosilicates

Cyclosilicates, ring silicates, have linked tetrahedra with $(Si_xO_{3x})^{2x-}$ or a ratio of 1:3. These exists as 3-member $(Si_3O_9)^6$, 4-member $(Si_4O_{12})^{8-}$ and 6-member $(Si_6O_{18})^{12-}$ rings.

3-member ring

- Benitoite - $BaTi(Si_3O_9)$

4-member ring

- Axinite - $(Ca,Fe,Mn)_3Al_2(BO_3)(Si_4O_{12})(OH)$

6-member ring

- Beryl/Emerald - $Be_3Al_2(Si_6O_{18})$
- Cordierite - $(Mg,Fe)_2Al_3(Si_5AlO_{18})$
- Tourmaline-$(Na,Ca)(Al,Li,Mg)_{3-}(Al,Fe,Mn)_6\,(Si_6O_{18})(BO_3)_3(OH)_4$

Inosilicates

Inosilicates, chain silicates, have interlocking chains of silicate tetrahedra with either SiO_3, 1:3 ratio, for single chains or Si_4O_{11}, 4:11 ratio, for double chains.

Single Chain Inosilicates:

Pyroxene group

- Enstatite - orthoferrosilite series
- Enstatite - $MgSiO_3$
- Ferrosilite - $FeSiO_3$
- Pigeonite - $CaO_{.25}(Mg,Fe)_{1.75}Si_2O_6$
- Diopside - hedenbergite series
- Diopside - $CaMgSi_2O_6$
- Hedenbergite - $CaFeSi_2O_6$
- Augite - $(Ca,Na)(Mg,Fe,Al)(Si,Al)_2O_6$
- Sodium pyroxene series
- Jadeite - $NaAlSi_2O_6$

- Aegirine (Acmite) - $NaFe^{3+}Si_2O_6$
- Spodumene - $LiAlSi_2O_6$

Pyroxenoid group

- Wollastonite - $CaSiO_3$
- Rhodonite - $MnSiO_3$
- Pectolite - $NaCa_2(Si_3O_8)(OH)$

Double Chain Inosilicates:

Amphibole group

- Anthophyllite - $(Mg,Fe)_7Si_8O_{22}(OH)_2$
- Cumingtonite series
- Cumingtonite - $Fe_2Mg_5Si_8O_{22}(OH)_2$
- Grunerite - $Fe_7Si_8O_{22}(OH)_2$
- Tremolite series
- Tremolite - $Ca_2Mg_5Si_8O_{22}(OH)_2$
- Actinolite - $Ca_2(Mg,Fe)_5Si_8O_{22}(OH)_2$
- Hornblende - $(Ca,Na)_{2\text{-}3}(Mg,Fe,Al)_5Si_6(Al,Si)_2O_{22}(OH)_2$

Sodium amphibole group

- Glaucophane - $Na_2Mg_3Al_2Si_8O_{22}(OH)$
- Riebeckite (asbestos) - $Na_2Fe^{2+}{}_3Fe^{3+}{}_2Si_8O_{22}(OH)_2$
- Arfvedsonite - $Na_3(Fe,Mg)_4FeSi_8O_{22}(OH)_2$

PHYLLOSILICATES

Phyllosilicates, sheet silicates, form parallel sheets of silicate tetrahedra with Si_2O_5 or a 2:5 ratio.

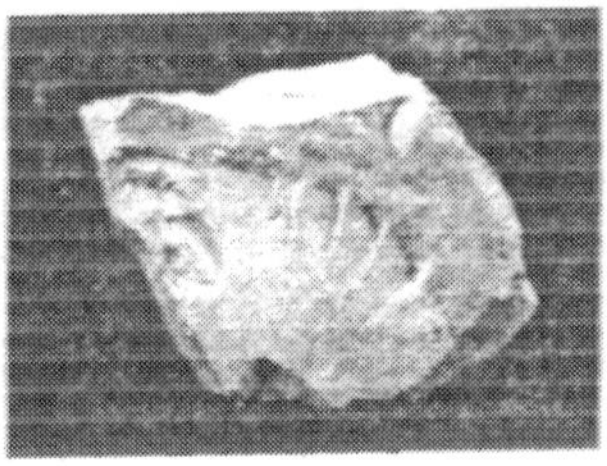

Kaolin

- Serpentine group
  - Antigorite - $Mg_3Si_2O_5(OH)_4$
  - Chrysotile - $Mg_3Si_2O_5(OH)_4$
  - Lizardite - $Mg_3Si_2O_5(OH)_4$
- Clay mineral group
  - Kaolinite - $Al_2Si_2O_5(OH)_4$
  - Talc - $Mg_3Si_4O_{10}(OH)_2$
  - Pyrophyllite - $Al_2Si_4O_{10}(OH)_2$
- Mica group
  - Muscovite - $KAl_2(AlSi_3O_{10})(OH)_2$
  - Phlogopite - $KMg_3Si_4O_{10}(OH)_2$
  - Biotite - $K(Mg,Fe)_3(AlSi_3O_{10})(OH)_2$
  - Lepidolite - $K(Li,Al)_{2-3}(AlSi_3O_{10})(OH)_2$
  - Margarite - $CaAl_2(Al_2Si_2O_{10})(OH)_2$
- Chlorite group
  - Chlorite - $(Mg,Fe)_3(Si,Al)_4O_{10}(OH)_2(Mg,Fe)3(OH)_6$

Tectosilicates

Tectosilicates, or "framework silicates", have a three-dimensional framework of silicate tetrahedra with SiO_2 or a 1:2 ratio. This is the largest group, comprising nearly 75 per cent of the crust of the Earth. Tectosilicates with the exception of the quartz group are aluminosilicates.

Quartz Crystal

- Quartz group
  - — Quartz - SiO_2
  - — Tridymite - SiO_2
  - — Cristobalite - SiO_2
  - — Opal - $SiO_2 \cdot nH_2O$
- Feldspar group
  - — Alkali-feldspars
  - — Potassium-feldspars
  - — Microcline - $KAlSi_3O_8$
  - — Orthoclase - $KAlSi_3O_8$
  - — Sanidine - $KAlSi_3O_8$
  - — Anorthoclase - $(Na,K)AlSi_3O_8$
  - — Albite - $NaAlSi_3O_8$
  - — Plagioclase feldspars
  - — Albite - $NaAlSi_3O_8$
  - — Oligoclase - $(Na,Ca)(Si,Al)_4O_8$ (Na:Ca 4:1)
  - — Andesine - $(Na,Ca)(Si,Al)_4O_8$ (Na:Ca 3:2)
  - — Labradorite - $(Na,Ca)(Si,Al)_4O_8$ (Na:Ca 2:3)
  - — Bytownite - $(Na,Ca)(Si,Al)_4O_8$ (Na:Ca 1:4)
  - — Anorthite - $CaAl_2Si_2O_8$
- Feldspathoid group
  - — Leucite - $KAlSi_2O_6$
  - — Nepheline - $(Na,K)AlSiO_4$
  - — Sodalite - $Na_8(AlSiO_4)_6Cl_2$
  - — Lazurite - $(Na,Ca)_8(AlSiO_4)_6(SO_4,S,Cl)_2$
  - — Petalite - $LiAlSi_4O_{10}$
- Scapolite group
  - — Marialite - $Na_4(AlSi_3O_8)_3(Cl_2,CO_3,SO_4)$
  - — Meionite - $Ca_4(Al_2Si_2O_8)_3(Cl_2CO_3,SO_4)$
  - — Analcime - $NaAlSi_2O_6 \cdot H_2O$

- Zeolite group
  - — Natrolite - $Na_2Al_2Si_3O_{10}$. $2H_2O$
  - — Chabazite - $CaAl_2Si_4O_{12}$. $6H_2O$
  - — Heulandite - $CaAl_2Si_7O_{18}$. $6H_2O$
  - — Stilbite - $NaCa_2Al_5Si_{13}O_{36}$. $17H_2O$

Carbonate Class

The carbonate minerals consist of those minerals containing the anion $(CO_3)^{2-}$ and include calcite and aragonite (both calcium carbonate), dolomite (magnesium/calcium carbonate) and siderite (iron carbonate). Carbonates are commonly deposited in marine settings when the shells of dead planktonic life settle and accumulate on the sea floor. Carbonates are also found in evaporitic settings (e.g. the Great Salt Lake, Utah) and also in karst regions, where the dissolution and reprecipitation of carbonates leads to the formation of caves, stalactites and stalagmites. The carbonate class also includes the nitrate and borate minerals.

Carbonate Minerals

Carbonate minerals are those minerals containing the carbonate ion: CO_3^{2-}.

Anhydrous Carbonates

- Calcite group: Trigonal
  - — Calcite $CaCO_3$
  - — Gaspeite $(Ni,Mg,Fe^{2+})CO_3$
  - — Magnesite $MgCO_3$
  - — Otavite $CdCO_3$
  - — Rhodochrosite $MnCO_3$
  - — Siderite $FeCO_3$
  - — Smithsonite $ZnCO_3$
  - — Sphaerocobaltite $CoCO_3$

- Aragonite group: Orthorhombic
  - Aragonite $CaCO_3$
  - Cerussite $PbCO_3$
  - Strontianite $SrCO_3$
  - Witherite $BaCO_3$
  - Rutherfordine UO_2CO_3
  - Natrite Na_2CO_3

Anhydrous Carbonates with Compound Formulas

- Dolomite group: Trigonal
  - Ankerite $CaFe(CO_3)_2$
  - Dolomite $CaMg(CO_3)_2$
  - Minrecordite $CaZn(CO_3)_2$
  - Barytocalcite $BaCa(CO_3)_2$

Carbonates with Hydroxyl or Halogen

- Carbonate with hydroxide: Monoclinic
  - Azurite $Cu_3(CO_3)_2(OH)_2$
  - Hydrocerussite $Pb_3(CO_3)_2(OH)_2$
  - Malachite $Cu_2CO_3(OH)_2$
  - Rosasite $(Cu,Zn)_2CO_3(OH)_2$
  - Phosgenite $Pb_2(CO_3)Cl_2$
  - Hydrozincite $Zn_5(CO_3)_2(OH)_6$
  - Aurichalcite $(Zn,Cu)_5(CO_3)_2(OH)_6$

Hydrated Carbonates

Ikaite $CaCO_3.6(H_2O)$

Lansfordite $MgCO_3 \cdot 5(H_2O)$

Monohydrocalcite $CaCO_3.H_2O$

Natron $Na_2CO_3 \cdot 10(H_2O)$

Zellerite $Ca(UO_2)(CO_3)_2 \cdot 5(H_2O)$

The carbonate class in both the Dana and the Srunz

classification systems includes the nitrates, the borates, and the iodates.

Sulphate Class

Sulphates all contain the sulphate anion, SO_4^{2-}. Sulphates commonly form in evaporitic settings where highly saline waters slowly evaporate, allowing the formation of both sulphates and halides at the water-sediment interface. Sulphates also occur in hydrothermal vein systems as gangue minerals along with sulphide ore minerals.

Another occurrence is as secondary oxidation products of original sulphide minerals. Common sulphates include anhydrite (calcium sulphate), celestine (strontium sulphate), barite (barium sulphate), and gypsum (hydrated calcium sulphate). The sulphate class also includes the chromate, molybdate, selenate, sulphite, tellurate, and tungstate minerals.

Anhydrite

Anhydrite

Anhydrite is a mineral - anhydrous calcium sulphate, $CaSO_4$. It is in the orthorhombic crystal system, with three directions of perfect cleavage parallel to the three planes of symmetry. It is not isomorphous with the orthorhombic barium (barite) and strontium (celestine) sulphates, as might be expected from the chemical formulas. Distinctly developed crystals are somewhat rare, the mineral usually presenting the form of cleavage masses. The hardness is 3.5 and the specific gravity 2.9. The colour is white, sometimes greyish, bluish or purple. On the best developed of the three cleavages the lustre is pearly, on other surfaces it is vitreous. When exposed to water, anhydrite readily transforms to the more commonly occurring gypsum, ($CaSO_4 \cdot 2H_2O$) by the absorption of water. Anhydrite is commonly associated with calcite, halite, and sulphides such as galena, chalcopyrite, molybdenite, and pyrite in vein deposits.

Anhydrite is most frequently found in salt deposits with gypsum; it was, for instance, first discovered, in 1794, in a salt mine near Hall in Tirol. In this occurrence depth is critical since nearer the surface anhydrite has been altered to gypsum by absorption of circulating ground water. From an aqueous solution calcium sulphate is deposited as crystals of gypsum, but when the solution contains an excess of sodium or potassium chloride anhydrite is deposited if temperature is above 40°C. This is one of the several methods by which the mineral has been prepared artificially, and is identical with its mode of origin in nature, the mineral is common in salt basins.

The name anhydrite was given by A. G. Werner in 1804, because of the absence of water of crystallisation, as contrasted with the presence of water in gypsum. Some obsolete names for the species are muriacite and karstenite; the former, an earlier name, being given under the impression that the substance was a chloride (muriate). A peculiar variety occurring as contorted concretionary masses is known as tripe-stone, and a scaly granular variety, from Vulpino, near Bergamo, in Lombardy, as vulpinite; the latter is cut and polished for ornamental purposes.

Chicken-wire Anhydrite

Chicken-wire anhydrite is a type of anhydrite deposit where seeping ground water gradually deposited large amounts of anhydrite in the sediment, replacing most of it, so that a cross-section of it looks somewhat like the coarse wire netting often used to confine poultry.

Halide Class

The halides are the group of minerals forming the natural salts and include fluorite (calcium fluoride), halite (sodium chloride), sylvite (potassium chloride), and sal ammoniac (ammonium chloride). Halides, like sulphates, are commonly found in evaporitic settings such as playa lakes and landlocked seas such as the Dead Sea and Great Salt Lake. The halide class includes the fluoride, chloride, and iodide minerals.

Fluorite

Fluorite (also called fluor-spar) is a mineral composed of calcium fluoride, CaF_2. It is an isometric mineral with a cubic habit, though octahedral and more complex isometric forms are not uncommon. Crystal twinning is common and adds complexity to the observed crystal habits.

Occurrence

Fluorite may occur as a vein deposit, especially with metallic minerals, where it often forms a part of the gangue (the worthless "host-rock" in which valuable minerals occur) and may be associated with galena, sphalerite, barite, quartz, and calcite. It is a common mineral in deposits of hydrothermal origin and has been noted as a primary mineral in granites and other igneous rocks and as a common minor constituent of dolostone and limestone.

Fluorite is a widely occurring mineral which is found in large deposits in many areas. Notable deposits occur in Germany, Austria, Switzerland, England, Norway, Mexico, and Ontario in Canada. In the United States deposits are found in Missouri, Oklahoma, Illinois, Kentucky, Colorado, New Mexico, Arizona, Ohio, New Hampshire, New York, and Texas. Illinois has historically been the largest producer of fluorite in the United States, however, the last of the mines closed in 1995. The Illinois general assembly passed a resolution in 1965 declaring Fluorite as the official state mineral.

FLUORESCENCE

Fluorite gives its name to the property of fluorescence, as many samples fluoresce strongly in ultra-violet light. The fluorescence may be due to impurities such as yttrium or organic matter in the crystal lattice.

Unfortunately not all fluorites fluoresce, even from the same locality, so UV lighting is not a reliable tool for identification of specimens, or for quantifying the mineral in mixtures. For example, among British fluorites, those from Northumberland, Durham and Eastern Cumbria are the most consistently fluorescent. Those from Yorkshire, Derbyshire and Cornwall are generally either not fluorescent, or feebly fluorescent.

Fluorite's fluorescence colour is largely dependent on where the original specimen was located. Blue is the most common colour but red, purple, yellow, green and white also occur. Fluorite also exhibits the property of thermoluminescence.

Blue John: One of the most famous of the older-known localities of fluorite is Castleton in Derbyshire, England, where, under the name of Derbyshire Blue John, beautiful purple-blue fluorite was extracted from the Blue John Cavern. It was used for ornamental purposes, especially in the 19th century. The name derives from French "bleu et jaune" (blue and yellow) characterising its colour. It is now scarce, and only a few hundred kilograms are mined each year for ornamental and lapidary use. Recent deposits in China have produced fluorite with similar colouring and banding to the classic Blue John stone.

Uses: As well as ornamental uses, fluorite is used in the making of opalescent glass, enamels cooking utensils, hydrofluoric acid, and as a flux in the manufacture of steel. Fluorite is also used in some high performance telescopes and camera lens elements instead of glass. Exposure tools for the semi-conductor industry, make the use of fluorite for the optics for 157 nm wavelength.

This wavelength is created by an excimer laser with F_2 gas, and the fluorite is a unique material that has high transparency at this wavelength. It has a very low dispersion so light diffraction is far less than ordinary glass, and in telescopes it allows crisp images of astronomical objects even at high power. The name fluorite is derived from the Latin *fluo,* meaning "flow", in reference to its use as a flux. Fluorite is slightly soluble in water, and is decomposed by sulphuric acid to form free hydrofluoric acid, which etches glass.

HALITE

Halite is the mineral form of sodium chloride, NaCl, commonly known as rock salt. Halite forms isometric crystals. The mineral is typically colourless to white, but may also be light blue, dark blue, and pink. It commonly occurs with other evaporite deposit minerals such as several of the sulphates, halides and borates.

Halite

Halite occurs in vast beds of sedimentary evaporite minerals that result from the drying up of enclosed lakes, playas, and seas. Salt beds may be up to 350 m thick and underlie broad areas. In the United States and Canada extensive underground beds extend from the Appalachian basin of western New York through parts of Ontario and under much of the Michigan basin. Other deposits are in Ohio, Kansas, New Mexico, Nova Scotia, and Saskatchewan.

Salt domes are vertical diapirs or pipe-like masses of salt that have been essentially "squeezed up" from underlying salt beds by mobilisation due to the weight of overlying rock. Salt domes contain anhydrite, gypsum, and native sulphur, in addition to halite and sylvite.

They are common along the Gulf coasts of Texas and Louisiana and are often associated with petroleum deposits. Germany, Spain, the Netherlands, Romania, and Iran also have salt domes. Salt *glaciers* exist in arid Iran where the salt has broken through the surface at high elevation and *flows* downhill. In all of these cases, halite is said to be behaving in the manner of a rhenium.

Unusual, purple, fibrous vein filling halite is found in France and a few other localities. Halite crystals termed *hopper crystals* appear to be "skeletons" of the typical cubes, with the edges present and stairstep depressions on, or rather in, each crystal face. In a rapidly crystallising environment the edges of the cubes simply grow faster than the centres.

Halite crystals form very quickly in some rapidly evaporating lakes resulting in modern artefacts with a coating or encrustation of halite crystals. *Halite flowers* are rare stalactites of curling fibres of halite that are found in certain arid caves of Australia's Nullarbor Plain. Halite stalactites and encrustations are also reported in the Quincy native copper mine of Hancock, Michigan.

Sylvite

Sylvite is potassium chloride (KCl) in natural mineral form. It forms crystals in the isometric system very similar to normal rock salt, halite (NaCl). (The two are, in fact, isomorphous. Sylvite is colourless to white with shades of yellow and red due to inclusions. It has a Mohs hardness of 2.5 and a specific gravity of 1.99. It has a refractive index of n=1.490) (Deer, et al.). Sylvite has a salty taste with a distinct bitterness.

Sylvite is one of the last evaporite minerals to precipitate out of solution. As such, it is only found in very dry saline areas. Its principle use is as a potassium fertilizer.

Sylvite is found in many evaporite deposits worldwide. Massive bedded deposits occur in New Mexico and western Texas, and in Utah in the US, but the largest world source is in Saskatchewan, Canada. The vast deposits in Saskatchewan, Canada were formed by the evaporation of a Devonian seaway. Sylvite is the official mineral of Saskatchewan.

Sylvite was first described in 1832 at Mt. Vesuvius near Napoli in Italy and named for the Dutch chemist, Francois Sylvius de le Boe (1614-1672).

Sal Ammoniac

Sal ammoniac is a rare mineral composed of ammonium chloride, NH_4Cl. It forms colourless to white to yellow brown crystals in the isometric - hexoctahedral class. It has very poor cleavage and a brittle to conchoidal fracture. It is quite soft, with a Mohs hardness of 1.5 to 2 and has a low specific gravity of 1.5. It is water soluble.

It typically forms as encrustations formed by sublimation around volcanic vents. It is found around volcanic fumaroles, guano deposits and burning coal seams. Associated minerals include: sodium alum, native sulphur and other fumarole minerals. Notable occurrences include: Tazikistan; Mt. Vesuvius, Italy and Paricutin, Michoacan, Mexico.

It is also the archaic name for the chemical compound ammonium chloride. It is commonly used in stained and leaded glass manufacture as a flux and tinning agent for irons used in the soldering process. In both jewellery making and the refining of precious metals, potassium carbonate (cream of tartar) is added to gold and silver in a borax-coated crucible to purify iron or steel filings that may have contaminated the scrap. It is then air cooled and remelted with a one-to-one mixture of powdered charcoal and sal ammoniac to yield a sturdy ingot of the respective metal or alloy in the case of sterling silver (.75 per cent copper) or karated gold. Anything other than 24-karat gold has silver and copper added.

Usually the addition of silica, zinc, and deoxidants in very small amounts relative to the pennyweight (dwt.) of gold are processed into gold from as low as 8kt to as high as 23.5 karat gold. This is added to prevent porosity and/or cracking while milling the ingot further into wire, sheet, or tubing. Without those additives an otherwise poor-quality ingot will result in open crucible melting with a hand torch or blowpipe and flame, as was done before electric melting furnaces were invented for use in the precious metals industry. These practices are still used by metalsmiths and jewellers today.

Sal ammoniac has also been used in the past in bakery products to give cookies a very crisp texture, although that application is rapidly dying due to the general disuse of it as an ingredient. The term fell out of general use largely in the 20th century.

Oxide Class

Oxides are extremely important in mining as they form many of the ores from which valuable metals can be extracted. They also carry the best record of changes in the Earth's magnetic field. They commonly occur as precipitates close to the Earth's surface, oxidation products of other minerals in the near surface weathering zone, and as accessory minerals in igneous rocks of the crust and mantle.

Common oxides include haematite (iron oxide), magnetite (iron oxide), chromite (iron chromium oxide), spinel (magnesium aluminium oxide - a common component of the mantle), ilmenite (iron titanium oxide), rutile (titanium dioxide), and ice (hydrogen oxide). The oxide class includes the oxide and the hydroxide minerals.

HAEMATITE

Haematite (AE) or haematite (BE) is the mineral form of Iron(III) oxide, (Fe_2O_3), one of several iron oxides. Haematite crystallises in the rhombohedral system, and it has the same crystal structure as ilmenite and as corundum. Haematite and ilmenite form a complete solid solution at temperatures above 950°C.

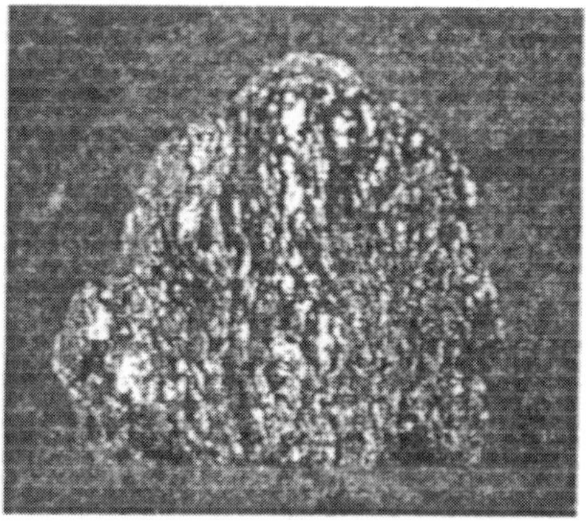

Haematite

Haematite is a very common mineral, coloured black to steel or silver-gray, brown to reddish brown, or red. It is mined as the main ore of iron. Varieties include *kidney ore, martite* (pseudomorphs after magnetite), *iron rose* and *specularite* (specular haematite). While the forms of haematite vary, they all have a rust-red streak. Haematite is harder than pure iron, but much more brittle.

Huge deposits of haematite are found in banded iron formations. Grey haematite is typically found in places where there has been standing water or mineral hot springs, such as those in Yellowstone. The mineral can precipitate out of water and collect in layers at the bottom of a lake, spring, or other standing water. Haematite can also occur without water, however, usually as the result of volcanic activity.

Clay-sized haematite crystals can also occur as a secondary mineral formed by weathering processes in soil, and along with other iron oxides or oxyhydroxides such as goethite, is responsible for the red colour of many tropical, ancient, or otherwise highly weathered soils.

The name haematite is derived from the Greek word for blood (haima) because haematite can be red, as in rouge, a powdered form of haematite. The colour of haematite lends it well in use as a pigment.

Especially good specimens of haematite come from England, Mexico, Brazil, Australia and the Lake Superior region of the United States and Canada.

Haematite is an anti-ferromagnetic material below the Morin transition at 260K, and weakly ferromagnetic above the Morin transition and below its Neel temperature at 948K, above which it is paramagnetic.

Haematite is part of a complex solid solution oxyhydroxide system having various degrees of water, hydroxyl group, and vacancy substitutions that affect the mineral's magnetic and crystal chemical properties. Two other end-members are referred to as protohaematite and hydrohaematite.

Jewellery

Polished haematite has been used in jewellery for at least 50 years in North America, especially in the western United States where it is found in jewellery and art created by the Native Americans. Its use in jewellery in Europe peaked in Victorian times when it was very popular.

Magnetite

Magnetite is a ferrimagnetic mineral with chemical formula Fe_3O_4, one of several iron oxides and a member of the spinel group. The chemical IUPAC name is iron (II,III) oxide and the common chemical name ferrous-ferric oxide.

The Curie temperature of magnetite is about 580°C. Magnetite is the most magnetic of all the naturally occurring minerals on Earth, and these magnetic properties led to lodestone being used as an early form of magnetic compass. Magnetite typically carries the dominant magnetic signature in rocks, and so it has been a critical tool in paleomagnetism, a science important in discovering and understanding plate tectonics. The relationships between magnetite and other iron-rich oxide minerals such as ilmenite, haematite, and ulvospinel have been much studied, as the complicated reactions between these minerals and oxygen influence how and when magnetite preserves records of the Earth's magnetic field.

Magnetite has been very important in understanding the conditions under which rocks form and evolve. Magnetite reacts with oxygen to produce haematite, and the mineral pair forms a buffer that can control oxygen fugacity. Commonly igneous rocks contain grains of two solid solutions, one between magnetite and ulvospinel and the other between ilmenite and haematite. Compositions of the mineral pairs are used to calculate how oxidising was the magma (i.e., the oxygen fugacity of the magma): a range of oxidising conditions are found in magmas and the oxidation state helps to determine how the magmas might evolve by fractional crystallisation.

Small grains of magnetite occur in almost all igneous rocks and metamorphic rocks. Magnetite also occurs in many sedimentary rocks, including banded iron formations. In many igneous rocks, magnetite-rich and ilmenite-rich grains occur that precipitated together from magma. Magnetite also is produced from peridotites and dunites by serpentinisation.

Magnetite is a valuable source of iron ore. It dissolves slowly in hydrochloric acid.

Distribution of Deposits

Magnetite is sometimes found in large quantities in beach sand. Such mineral sands or iron sands or black sands are found in various places such as California and the west coast of New Zealand. The magnetite is carried to the beach via rivers from erosion and is concentrated via wave action and currents.

Huge deposits have been found in banded iron formations. These sedimentary rocks have been used to infer changes in the oxygen content of the atmosphere of the Earth.

Large deposits of magnetite also are found in Kiruna, Sweden, the Pilbara region in Western Australia, and in the Adirondack region of New York in the United States. Deposits are also found in Norway, Germany, Italy, Switzerland, South Africa, India, Mexico, and in Oregon, New Jersey, Pennsylvania, North Carolina, Virginia, New Mexico, Utah, and Colorado in the United States. Recently, in June 2005, an exploration company, Cardero Resources, discovered a vast deposit of magnetite-bearing sand dunes in Peru. The dune field covers 250 km^2, with the highest dune at over 2000 m above the desert floor. The sand contains 10 per cent magnetite.

Biological Occurrences

Crystals of magnetite have been found in some bacteria (e.g., Magnetospirillum magnetotacticum) and in the brains of bees, of termites, of some birds (e.g., the pigeon), and of humans. These crystals are thought to be involved in magnetoreception, the ability to sense the polarity or the inclination of the earth's magnetic field, and to be involved in navigation. Also, chitons have teeth made of magnetite on their radula making them unique among animals. This means they have an exceptionally abrasive tongue with which to scrape food from rocks.

The study of biomagnetism began with the discoveries of Caltech paleoecologist Heinz Lowenstam in the 1960s.

Preparation as a Ferrofluid

Magnetite can be prepared in the laboratory as a ferrofluid in the Massart method by mixing iron (II) chloride and iron (III) chloride in the presence of sodium hydroxide.

CHROMITE

Chromite, iron magnesium chromium oxide: (Fe, Mg)Cr_2O_4, is an oxide mineral belonging to the spinel group. Magnesium is always present in variable amounts, also aluminium and iron substitute for chromium.

Chromite is found in peridotite and other layered ultramafic intrusive rocks and also found in metamorphic rocks such as serpentinites. Ore deposits of chromite form as early magmatic differentiates. It is commonly associated with olivine, magnetite, serpentine, and corundum. The vast Bushveld igneous complex of South Africa is a large layered mafic to ultramafic igneous body with some layers consisting of 90 per cent chromite making the rare rock type, chromitite. Chromite is also used as a refractory material.

The only ore of chromium is the mineral chromite. In the western hemisphere, chromite ore is produced only in Brazil and Cuba; By comparison, about 80 per cent of world production of chromite comes from India, Iran, Pakistan, Oman, Zimbabwe, Turkey and Southern Africa. Southern Africa itself produces about half of this.

Chromite is mined from the ultramafic rocks in the Zhob District of Balochistan, Most of the chromite is of metallurgical grade with Cr_2O_3 averaging 46 per cent and a chrome to iron ratio of 3:1.

Spinel

The spinels are any of a class of minerals which crystallise in the isometric system with an octahedral habit. The general formula is as $(X)(Y)_2O_4$, with X representing cations occupying tetrahedral sites and Y cations occupying octahedral sites. Divalent, trivalent, and quadrivalent cations can occupy the X and Y sites, and they include magnesium, zinc, iron, manganese, aluminium, chromium, titanium, and silicon.

The oxygen anions are arranged in a cubic close-packed structure. In the normal spinel structure, X cations occupy the tetrahedral sites, and Y the octahedral sites. For inverse spinels, half the Y cations occupy the tetrahedral sites, and both X and Y cations occupy the octahedral sites.

Important members of the spinel group include:

- Spinel – $MgAl_2O_4$, after which this class of minerals is named
- Gahnite - $ZnAl_2O_4$
- Franklinite - $(Fe,Mn,Zn)(Fe,Mn)_2O_4$
- Chromite - $(Fe{\cdot}Mg)Cr_2O_4$
- Magnetite - Fe_3O_4
- Hercynite - $FeAl_2O_4$
- Ulvospinel - $TiFe_2O_4$
- Jacobsite - $MnFe_2O_4$
- Trevorite - $NiFe_2O_4$

Ringwoodite - $SiMg_2O_4$, an abundant olivine polymorph within the Earth's mantle from about 520 to 660 km depth, and a rare mineral in meteorites.

True spinel has long been found in the gemstone-bearing gravel of Sri Lanka and in limestones of Myanmar and Thailand.

Spinel usually occurs in isometric crystals, octahedrons, usually twinned. It has an imperfect octahedral cleavage and a conchoidal fracture. Its hardness is 8, its specific gravity is 3.5-4.1 and it is transparent to opaque with a vitreous to dull lustre. It may be colourless, but is usually various shades of red, blue, green, yellow, brown or black. There is a unique natural white spinel, now lost, that surfaced briefly in what is now Sri Lanka. Another famous spinel is the Black Prince's Ruby in the British Crown Jewels.

The transparent red spinels are called spinel-rubies or balas-rubies and were often confused with actual rubies in ancient times. "Balas" is derived from Balascia, the ancient name for Badakhshan, a region in central Asia situated in the upper valley of the Kokcha river, one of the principal tributaries of the Oxus river. Yellow spinel is called rubicelle and violet-coloured manganese-bearing spinel is called almandine.

Spinel is found as a metamorphic mineral, and also as a

primary mineral in basic rocks, because in such magmas the absence of alkalis prevents the formation of feldspars and any aluminium oxide present will form corundum or combine with magnesia to form spinel. This is why spinel and ruby are often found together.

Spinel, $(Mg,Fe)(Al,Cr)_2O_4$, is common in peridotite in the uppermost Earth's mantle, between the Mohorovicic discontinuity (the Moho) and a depth of 70 kilometres or so; below that depth, the spinel (if present) becomes increasingly rich in chromium, as with increasing depth, pyrope-rich garnet becomes the more stable aluminous mineral in peridotite. At depths significantly shallower than the Moho, calcic plagioclase is the more stable aluminous mineral in peridotite.

Ilmenite

Ilmenite is a weakly magnetic iron-black or steel-gray mineral found in metamorphic and igneous rocks. It is a crystalline iron titanium oxide ($FeTiO_3$). It crystallises in the rhombohedral system, and it has the same crystal structure as corundum and haematite. Many mafic igneous rocks contain grains of intergrown magnetite and ilmenite, formed by the oxidation of ulvospinel. Ilmenite also occurs as discrete grains, typically with some haematite in solid solution, and complete solid solution exists between the two minerals at temperatures above about 950°C. Ilmenite also forms solid solutions with geikielite ($MgTiO_3$).

Titanium was identified for the first time by William Gregor in 1791 in Ilmenite from the Manaccan valley.

Sources

Although named after the locality of its discovery in the Il'menski Mountains, near Miass, Russia, most ilmenite is recovered from secondary sources, such as beach sands. It is normally found in sand as rounded particles with a diameter between 0.1 and 0.2 mm. Fraser Island and Rainbow Beach, Australia show enormous natural abundance of ilmenite and other minerals and had to be protected through high court to establish the Great Sandy National Park.

The world's largest opencast ilmenite mine is the Tellnes mine located in Sokndal, Norway and run by Titania AS (owned by Kronos Worldwide Inc.). In Karhujupukka located in Kolari, northern Finland there is a magnetite - ilmenite ore at around 5 million tons. The ore contains about 6.2 per cent titanium.

The majority of the ilmenite mined is used as a raw material for pigment production. The product is titanium dioxide, which is an extremely white substance used as a base in high-quality paint. It is commonly massive, but also occurs in rhombohedral crystals. There are large mineral sands deposits within South Africa as well, one such deposit is to be found at Richard's Bay.

Lunar Ilmenite

Ilmenite has been found in Moon rocks, and in 2005 NASA used the Hubble Space Telescope to locate potentially ilmenite-rich locations. This mineral could be essential to an eventual Moon base, as ilmenite would provide a source of iron and titanium for the building of structures and essential oxygen extraction.

RUTILE

Rutile needles

Rutile is a mineral composed primarily of titanium dioxide, TiO_2.

Rutile is the most common natural form of TiO_2, with two rarer polymorphs anatase (sometimes known by the obsolete name 'octahedrite'), a tetragonal mineral of pseudo-octahedral habit; and brookite, an orthorhombic mineral.

Rutile has among the highest refractive indices of any known mineral and also exhibits high dispersion. Natural rutile may contain up to 10 per cent iron and significant amounts of niobium and tantalum.

Rutile derives its name from the Latin *rutilus*, red, in reference to the deep red colour observed in some specimens when viewed by transmitted light.

Occurrence

Rutile is a common accessory mineral in high-temperature and high-pressure metamorphic rocks and in igneous rocks.

Rutile is the preferred polymorph of TiO_2 in such environments because it has the lowest molecular volume of the three polymorphs; it is thus the primary titanium bearing phase in most high pressure metamorphic rocks, chiefly eclogites. Brookite and anatase are typical polymorphs of rutile formed by retrogression of metamorphic rutile.

Within the igneous environment, rutile is a common accessory mineral in plutonic igneous rocks, although it is also found occasionally in extrusive igneous rocks, particularly those which have deep mantle sources such as kimberlites and lamproites. Anatase and brookite are found in the igneous environment particularly as products autogenic alteration during the cooling of plutonic rocks; anatase is also found formed within placer deposits sourced from primary rutile.

The occurrence of large specimen crystals is most common in pegmatites, skarns and particularly granite greisens.

Rutile is found as an accessory mineral in some altered igneous rocks, and in certain gneisses and schists. In groups of acicular crystals it is frequently seen penetrating quartz as in the "fléches d'amour" from Grisons, Switzerland.

Uses and Economic Importance: Rutile, when present in large enough quantities in beach sands, forms an important constituent of heavy mineral sands ore deposits. It is primarily extracted for use in refractory manufacture or use as a base for paints. Rarely is it extracted as an ore of titanium.

Finely powdered rutile is a brilliant white pigment and is used in paints, plastics, papers, foods, and other applications that call for a bright white colour. Titanium dioxide pigment is the single greatest use of titanium worldwide. Nanoscale particles of rutile are transparent to optical light but remain highly reflective to UV light. Hence, they are used in sunscreens.

Small rutile needles present in gems are responsible for an optical phenomenon known as asterism. Asterated gems are known as "star" gems. Star sapphires, star rubies, and other "star" gems are highly sought after and often more valuable than their normal equivalents.

Synthetic Rutile: Synthetic rutile was first produced in 1948 and is sold under a variety of names. Very pure synthetic rutile is transparent and almost colourless (slightly yellow) in large pieces. Synthetic rutile can be made in a variety of colours by doping, although the purest material is almost colourless. The high refractive index gives an adamantine lustre and strong refraction that leads to a diamond-like appearance. The near-colourless diamond substitute is sold under the name Titania, which is the old-fashioned chemical name for this oxide. However, rutile is seldom used in jewellery because it is not very hard (scratch-resistant), measuring only about 6 on the Mohs hardness scale.

Sulphide Class

Many sulphide minerals are economically important as metal ores. Common sulphides include pyrite (iron sulphide-commonly known as fools' gold), chalcopyrite (copper iron sulphide), pentlandite (nickel iron sulphide), and galena (lead sulphide). The sulphide class also includes the selenides, the tellurides, the arsenides, the antimonides, the bismuthinides, and the sulphosalts (sulphur and a second anion such as arsenic). A sulphide mineral is a mineral containing sulphide (S^{2-}) as the major anion. Closely related and often included within the sulphide class are selenide and telluride minerals.

Pyrite

Pyrite fool'sgold

The mineral pyrite, or iron pyrite, is iron sulphide, FeS_2. It has isometric crystals that usually appear as cubes. The cube faces may be striated (parallel lines on crystal surface or cleavage face) as a result of alternation of the cube and pyritohedron faces. Pyrite also frequently occurs as octahedral crystals and as pyritohedra (a dodecahedron with pentagonal faces). It has a slightly uneven and conchoidal fracture, a hardness of 6–6.5, and a specific gravity of 4.95–5.10. It is brittle, meaning it breaks or powders easily. A major identifier, useful in the field, is the streak, as the powdered mineral smells of sulphur. Its metallic lustre and pale-to-normal, brass-yellow hue have earned it the nickname fool's gold due to many miners mistaking it for the real thing. Ironically, small quantities of actual gold are sometimes found in pyrite. In fact, such auriferous pyrite is a valuable ore of gold.

Pyrite is the most common of the sulphide minerals. It is usually found associated with other sulphides or oxides in quartz veins, sedimentary rock and metamorphic rock, as well as in coal beds, and as the replacement mineral in fossils.

Cubicpyrite

Pyrite exposed to the environment during mining and excavation reacts with oxygen and water to form sulphuric acid, resulting in acid mine drainage. This drainage results from the action of *Thiobacillus* bacteria, which generate their energy by using oxygen to oxidise ferrous iron (Fe^{2+}) to ferric iron (Fe^{3+}). The ferric iron in turn reacts with pyrite to produce ferrous iron and sulphuric acid. The ferrous iron is then available for oxidation by the bacteria; this cycle can continue until the pyrite is exhausted. Pyrite is used commercially for the production of sulphur dioxide, for use in such application as the paper industry, and in the manufacture of sulphuric acid, though such applications are declining in importance.

Pyrites can show negative resistance and have experimentally been used in oscillator circuits as radio detectors.

The name pyrite is from the Greek word (*pura*) meaning "fire". This is likely due to the sparks that result when pyrite is struck against steel. This capacity made it popular for use in early firearms such as the wheellock.

Cubes of fully or partially oxidised and hydrated pseudomorphs of pyrite are known colloquially as *devil's dice*.

Pyrite and Marcasite: Pyrite is often confused with the mineral marcasite, a name derived from the Arabic word for pyrite, due to their similar characteristics. Marcasite is a polymorph of pyrite, which means it has the same formula as pyrite but a different structure and, therefore, different symmetry and crystal shapes. The formal oxidation states are, however, the same as in pyrite because again the sulphur atoms occur in persulphide-like pairs. Marcasite/pyrite is probably the most famous polymorph pair next to the diamond/graphite pair.

Marcasite is metastable relative to pyrite and will slowly be changed to pyrite if heated or given enough time. Marcasite is relatively rare, but may be locally abundant in some types of ore deposits, such as Mississippi Valley-type Pb-Zn deposits. Marcasite appears to form only from aqueous solutions.

Pyrite is often used in jewellery such as necklaces and bracelets, but although the two are similar, marcasite cannot be used in jewellery as it has a tendency to crumble into powder. Adding to the confusion between marcasite and pyrite is the use of the word marcasite as a jewellery trade name. The term is applied to small polished and faceted stones that are inlaid in sterling silver, but even though they are called marcasite, they are actually pyrite.

Formal Oxidation States for Pyrite, Marcasite, and Arsenopyrite: From the perspective of classical inorganic chemistry, which assigns formal oxidation states to each atom, pyrite is probably best described as $Fe^{2+}S_2^{2-}$. This formalism recognises that the sulphur atoms in pyrite occur in pairs with clear S-S bonds. These persulphide units can be viewed as derived from hydrogen persulphide, H_2S_2. Thus pyrite would be more descriptively called iron persulphide, not iron disulphide. In contrast, molybdenite, MoS_2, features isolated sulphide (S^{2-}) centres. Consequently, the oxidation state of molybdenum is Mo^{4+}, or Mo(IV). The mineral arsenopyrite has the formula FeAsS. Whereas pyrite has S_2 subunits, arsenopyrite has AsS units, formally derived from deprotonation of H_2AsSH.

Analysis of classical oxidation states would recommend the description of arsenopyrite as $Fe^{3+}AsS^{3-}$. Of course these formalisms ignore covalency, which is strongly implied by the semi-conducting behaviour of this family of inorganic solids.

Related Minerals and Varieties: Bravoite is a nickeloan variety of pyrite, with >50 per cent substitution of Ni^{2+} for Fe^{2+} within pyrite. Bravoite is not a formally recognised mineral, and is named after Peruvian scientist, Jose J. Bravo (1874-1928).

CHALCOPYRITE

Chalcopyrite is a copper iron sulphide mineral that crystallises in the tetragonal system. It has the chemical composition $CuFeS_2$.

It has a brassy to golden yellow colour and a hardness of 3.5 to 4 on the Mohs scale. Its streak is diagnostic as green tinged black. The name is pronounced *kal-co-pie-right* (kal as in "*calendar*", "co" as in *co*at).

On exposure to air, chalcopyrite oxidises to a variety of oxides, hydroxides and sulphates. Associated copper minerals include the sulphides bornite (Cu_5FeS_4), chalcocite (Cu_2S), covellite (CuS), digenite (Cu_9S_5); carbonates such as malachite and azurite, and rarely oxides such as cuprite (Cu_2O). Chalcopyrite is rarely found in association with native copper.

Min Chalcopyrite

Identification: Chalcopyrite is often confused with pyrite, although the latter has a cubic and not a tetragonal crystal system. Further, chalcopyrite is often massive, rarely crystalline, and less brittle. Chalcopyrite is also a darker yellow in colour, with a greenish tinge and diagnostic greasy lustre.

Due to its colour and high copper content, chalcopyrite has often been referred to as "yellow copper".

Chemistry

Natural chalcopyrite has no solid solution series with any other sulphide minerals. There is limited substitution of Zn with Cu despite chalcopyrite having the same crystal structure as sphalerite. However, it is often contaminated by a variety of other trace elements such as Co, Ni, Mn, Zn and Sn substituting for Cu and Fe. Se, Fe and As substitute for sulphur, and trace amounts of Ag, Au, Pt, Pd, Pb, V, Cr, In, Al and Sb are reported.

It is likely many of these elements are present in finely intergrown minerals within the chalcopyrite crystal, for instance lamellae of arsenopyrite representing As, molybdenite representing Mo, etc.

Paragenesis

Chalcopyrite is present within many ore bearing environments via a variety of ore forming processes.

Chalcopyrite is present in volcanogenic massive sulphide ore deposits and sedimentary exhalative deposits, formed by deposition of copper during hydrothermal circulation. Chalcopyrite is concentrated in this environment via fluid transport.

Porphyry copper ore deposits are formed by concentration of copper within a granite stock during the ascent and crystallisation of a magma. Chalcopyrite in this environment is produced by concentration within a magma system.

Chalcopyrite is an accessory mineral in Kambalda type komatiitic nickel ore deposits, formed from an immiscible sulphide liquid in sulphur-saturated ultramafic lavas. In this environment chalcopyrite is formed by a sulphide liquid stripping copper from an immiscible silicate liquid.

Occurrence

Chalcopyrite is the most important copper ore. Chalcopyrite ore occurs in a variety of ore types, from huge masses as at Timmins, Ontario, to irregular veins and disseminations associated with granitic to dioritic intrusives as in the porphyry copper deposits of Broken Hill, the American cordillera and the Andes.

Chalcopyrite is present in the supergiant Olympic Dam Cu-Au-U deposit in South Australia.

Chalcopyrite may also be found in coal seams associated with pyrite nodules, and as disseminations in carbonate sedimentary rocks.

PENTLANDITE

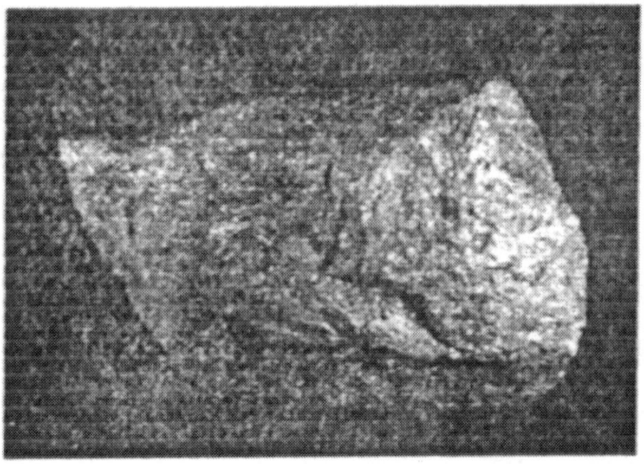

Pentlandite

Pentlandite is an iron-nickel sulphide, $(Fe,Ni)_9S_8$. Pentlandite usually has a Ni:Fe ratio of close to 1:1. It also contains minor cobalt. It is named after the Irish scientist Joseph Barclay Pentland (1797-1873), who first noted the mineral.

Pentlandite forms isometric crystals, but is normally found

in massive granular aggregates. It is brittle with a hardness of 3.5 - 4 and specific gravity of 4.6 - 5.0 and is non-magnetic. It has a yellowish bronze colour.

Paragenesis

Pentlandite is the most common terrestrial nickel sulphide mineral formed from immiscible sulphide-silicate melts under normal mantle and crustal conditions.

Nickel, being a chalcophile element, prefers to inhabit a sulphide phase versus a silicate or oxide phase within most terrestrial lithochemical systems (a few exceptions exist in unusual compositions). This behaviour is seen only when the particular rock is molten and sulphur saturated.

In sulphur undersaturated melts, nickel will substitute for other transition metals within ferromagnesian minerals, the most usual being olivine, although nickeliferous varieties of amphibole, biotite, pyroxene and spinels are known. Ni substitutes most readily for Fe^{2+} and Mg^{2+}.

In sulphur saturated melts, nickel behaves as a chalcophile element and partitions strongly into the sulphide phase. Because most nickel exists in ultramafic rocks and behaves as a compatible element in igneous differentiation processes, the formation of nickel-bearing sulphides is essentially restricted to sulphur saturated mafic and ultramafic melts.

The sulphide melt, being at or above 1000°C, is in the form of monosulphide solid solution (MSS), an amalgam of compositional "mineral" components of pentlandite, pyrite and pyrrhotite, and usually containing a small percentage of chalcopyrite (Cu being chalcophile), all of which are in an amorphous form. It is only upon cooling past ~550°C (dependent on composition) that the MSS undergoes exsolution into its component sulphide phases.

These phases are typically formed in an aphanitic equigranular granoblastic massive sulphide phase, or as matrix ore or disseminated sulphides held within the overlying silicate rock matrix. Intact magmatic massive sulphide is rarely preserved as, aside from the Norilsk deposit, most deposits of nickeliferous sulphide have been metamorphosed.

Metamorphism, especially if it is of at least middle greenschist facies, will cause the solid massive sulphide to revert to MSS. During deformation the MSS will act in a ductile fashion, and it is often considered to have the consistency of toothpaste, able to travel great distances into the country rock and along structures. Upon cessation of metamorphism, the MSS solution reverts again to the component sulphides, but it usually inherits a foliated or sheared texture, and typically sees growth of bright, equigranular to globular aggregates of porphyroblastic pentlandite crystals known colloquially as "fish scales".

Metamorphism may also reconstitute the MSS and sulphide composition, which may alter the concentration of Ni and the Ni:Fe ratio and Ni:S ratio of the sulphides. In this case, pentlandite may be replaced by millerite, and rarely heazlewoodite. Metamorphism may also see the introduction of aggressive metasomatism, and it is particularly common for arsenic to enter the MSS, producing niccolite, gersdorffite and other Ni-Co arsenides.

Occurrence

Pentlandite is found within the lower margins of mineralised ultramafic to mafic layered intrusions, the prime example being the Bushveld igneous complex, South Africa, the Voiseys Bay troctolite intrusive complex in Canada, the Duluth gabbro, in North America, and various other localities throughout the world. In these locations it forms an important nickel ore.

Pentlandite is also the principal ore mineral won from Kambalda type komatiitic nickel ore deposits, the type examples of which are in the Yilgarn Craton of Western Australia. Similar deposits exist at Nkomati, Namibia, in the Thompson Nickel Belt, Canada, and a few examples from Brazil.

Pentlandite, but primarily chalcopyrite and PGEs, are won from the supergiant Norilsk nickel deposit, in trans-Siberian Russia.

The Sudbury deposit in Ontario, Canada, is associated with a meteorite impact crater. Pentlandite-pyrite-pyrrhotite ore in this location was formed from an extensive melt sheet formed by melting of rock post- impact, which became sulphur saturated and formed extensive sheetlike ore deposits.

Phosphate Class

The phosphate mineral group actually includes any mineral with a tetrahedral unit AO_4 where A can be phosphorus, antimony, arsenic or vanadium. By far the most common phosphate is apatite which is an important biological mineral found in teeth and bones of many animals. The phosphate class includes the phosphate, arsenate, vanadate, and antimonate minerals.

Phosphate minerals are those minerals that contain the tetrahedrally coordinated phosphate (PO_4^{3-}) anion along with the freely substituting arsenate (AsO_4^{3-}) and vanadate (VO_4^{3}). Chlorine (Cl^-), fluorine (F^-), and hydroxide (OH^-) anions also fit into the crystal structure.

The phosphate class of minerals is a large and diverse group, however, only a few species are relatively common.

Examples Include:

- Triphylite $Li(Fe,Mn)PO_4$
- Monazite $(Ce,La,Y,Th)PO_4$
- Apatite group $Ca_5(PO_4)_3(F,Cl,OH)$
  — hydroxylapatite $Ca_5(PO_4)_3OH$
  — fluorapatite $Ca_5(PO_4)_3F$
  — chlorapatite $Ca_5(PO_4)_3Cl$
- Pyromorphite $Pb_5(PO_4)_3Cl$
- Vanadinite $Pb_5(VO_4)_3Cl$

- Erythrite $Co_3(AsO_4)_2 \cdot 8H_2O$
- Amblygonite $LiAlPO_4F$
- Lazulite $(Mg,Fe)Al_2(PO_4)_2(OH)_2$
- Wavellite $Al_3(PO_4)_2(OH)_3 \cdot 5H_2O$
- Turquoise $CuAl_6(PO_4)_4(OH)_8 \cdot 5H_2O$
- Autunite $Ca(UO_2)_2(PO_4)_2 \cdot 10\text{-}12H_2O$
- Carnotite $K_2(UO_2)_2(VO_4)_2 \cdot 3H_2O$
- Phosphophyllite $Zn_2(Fe,Mn)(PO_4)_2 \cdot 4H_2O$

Applications

Phosphate minerals are often used for control of rust and prevention of corrosion on ferrous materials, applied with electrochemical conversion coatings.

APATITE

Apatite is a group of phosphate minerals, usually referring to hydroxylapatite, fluorapatite, and chlorapatite, named for high concentrations of OH^-, F^-, or Cl^- ions, respectively, in the crystal. The formula of the admixture of the three most common species is written as $Ca_5(PO_4)_3(OH, F, Cl)$, and the formulae of the individual minerals are written as $Ca_5(PO_4)_3(OH)$, $Ca_5(PO_4)_3F$ and $Ca_5(PO_4)_3Cl$, respectively.

Apatite is one of few minerals that are produced and used by biological micro-environmental systems. Hydroxylapatite is the major component of tooth enamel. A relatively unique form of apatite in which most of the OH groups are absent and containing many carbonate and acid phosphate substitutions is a large component of bone material.

Fluorapatite (or fluoroapatite) is more resistant to acid attack that is hydroxyapatite. For this reason, toothpaste typically contain a source of fluoride anions (e.g. sodium fluoride, sodium monofluorophosphate). Similarly, fluoridated water, allow exchange in the teeth of fluoride ions for hydroxy groups in apatite. Too much fluoride results in dental fluorosis and/or skeletal fluorosis.

In the United States, apatite is often used to fertilize tobacco. It partially starves the plant of nitrogen, which gives American cigarettes a different taste from those of other countries.

Fission tracks in apatite are commonly used to determine the thermal history of orogenic (mountain) belts and of sediments in sedimentary basins. Phosphorite is the name given to impure, massive apatite.

Gemology

Apatite is infrequently used as a gemstone. Transparent stones of clean colour have been faceted, and chatoyant specimens have been cabochon cut. Chatoyant stones are known as *cat's-eye apatite,* transparent green stones are known as *asparagus stone,* and blue stones have been called *moroxite.* Crystals of rutile my have grown in the crystal of apatite so when in the right light, the cut stone displays a cat's eye effect. Major sources for gem apatite are: Brazil, Burma, and Mexico. Other sources include: Canada, Czechoslovakia, Germany, India, Madagascar, Mozambique, Norway, South Africa, Spain, Sri Lanka, and the US.

Vanadium (V) Oxide

Vanadium (V) oxide (V_2O_5), commonly known as vanadium pentoxide, is the most important compound of vanadium. Upon heating it can reversibly lose oxygen to the air. Related to this ability, V_2O_5 catalyses the aerobic oxidation of sulphur dioxide, benzene and naphthalene, which is the basis for its industrial production of sulphuric acid, maleic anhydrite, and phthalic anhydrite, respectively. It is a poisonous orange solid which, because of its high oxidation state, is both an amphoteric oxide and an oxidising agent. Unlike most metal oxides, it is slightly soluble in water.

Chemical Properties

In this compound, like all others having the vanadium (V) designation, the vanadium is in the +5 oxidation state. All of the oxygen atoms in the compound are in the -2 oxidation state. Vanadium (V) oxide dissolves slightly in water to give an acidic solution, though it is an amphoteric oxide. Thus it reacts with strong non-reducing acids to form solutions containing the pale yellow dioxovanadium (V) ion:

$$V_2O_5 + 2\ HNO_3 \rightarrow 2\ VO_2(NO_3) + H_2O$$

Thionyl chloride converts it to $VOCl_3$:

$$V_2O_5(s) + 3\ SOCl_2(l) \rightarrow 2\ VOCl_3(l) + 3\ SO_2(g)$$

It also reacts with strong alkali to form polyoxovanadates, which have a complex structure that depends on pH. If excess aqueous sodium hydroxide is used, the product is a colourless salt, sodium metavanadate, Na_3VO_4. If acid is slowly added to a solution of Na_3VO_4, the colour gradually deepens through orange to red before brown hydrated V_2O_5 precipitates around pH 2. These solutions contain mainly the ions HVO_4^{-2} and $V_2O_7^{4-}$ between pH 9 and 13, but below pH 9 more exotic species such as $V_4O_{12}^{4-}$ and $HV_{10}O_{28}^{5-}$ predominate.

V_2O_5 is easily reduced in acidic media to the stable vanadium (IV) species, the blue vanadyl ion ($VO(H_2O)_5^{2+}$). This conversion illustrates the redox properties of V_2O_5. For example, hydrochloric acid and hydrobromic acid are oxidised to the corresponding halogen, e.g.,

$$V_2O_5(s) + 6\ HCl + 7\ H_2O \rightarrow 2\ [VO(H_2O)_5]^{2+} + 4\ Cl^- + Cl_2$$

Solid V_2O_5 is reduced by oxalic acid, CO, and SO_2 to give vanadium (IV) oxide, VO_2 as a deep-blue solid. Further reduction using hydrogen or excess CO can lead to complex mixtures of oxides such as V_4O_7 and V_5O_9 before black V_2O_3 is reached. Vanadates or vanadyl (V) compounds in acid solution are reduced by zinc amalgam through the interestingly colourful pathway -

yellow VO_3^- and $VO_2^+ \rightarrow$ blue $VO^{2+} \rightarrow$ green $V^{3+} \rightarrow$ purple V^{2+}

Preparation: Technical grade V_2O_5 is produced as a black powder used for the production of vanadium metal and ferrovanadium. A vanadium ore or vanadium-rich residue is treated with sodium carbonate to prôduce sodium metavanadate, $NaVO_3$. This is then acidified to pH 2-3 using H_2SO_4 to yield a precipitate of "red cake". The red cake is then melted at 690°C to produce the crude V_2O_5.

Vanadium (V) oxide is also the main product when vanadium metal is heated with excess oxygen, but this product is contaminated with other lower oxides. A more satisfactory laboratory preparation involves the decomposition of ammonium metavanadate at around 200 °C:

$$2\ NH_4VO_3 \rightarrow V_2O_5(s) + 2\ NH_3 + H_2O$$

Uses: The most important use of vanadium (V) oxide is in the manufacture of sulphuric acid, an important industrial chemical with an annual production of 145 million tonnes in 1986. Vanadium (V) serves the crucial purpose of catalysing the mildly exothermic oxidation of sulphur dioxide to sulphur trioxide by air in the contact process:

$$2\ SO_2 + O_2 \rightarrow 2\ SO_3$$

The discovery of this simple reaction, for which V_2C_5 is the most effective catalyst, allowed sulphuric acid to become the cheap commodity chemical it is today. The reaction is performed between 400 and 620 °C; below 400 °C the V_2O_5 is inactive as a catalyst, and above 620 °C; it begins to break down. Since it is known that V_2O_5 can be reduced to VO_2 by SO_2, one likely catalytic cycle is as follows:

$$SO_2 + V_2O_5(s) \rightarrow SO_3(g) + 2\ VO_2(s) \text{ followed by}$$

$$2\ VO_2(s) + {}^1/_2\ O_2(g) \rightarrow V_2O_5$$

Paradoxically, it is also used as catalyst in the selective catalytic reduction of NO_x emissions in some power plants. Due to its effectiveness in converting sulphur dioxide into sulphur trioxide, and thereby sulphuric acid, special care must be taken with the operating temperatures and placement of a power plant's SCR unit when firing sulphur-containing fuels.

Maleic anhydrite is another important industrial material, used for the manufacture of polyester resins and alkide resins. Vanadium (V) oxide can catalyse its production from a variety of organic starting materials such as *n*-butane, furfural and benzene, the last of which is the usual commercial method. In a related process, phthalic anhydrite, used for making plasticisers for PVC manufacture, may be produced by V_2O_5 catalysed oxidation of ortho-xylene or naphthalene at 350-400°C.

In terms of quantity, the major use for vanadium (V) oxide is in the production of ferrovanadium. The oxide is heated with scrap iron and aluminium, producing the iron-vanadium alloy along with alumina as a by-product. In 2005 a shortage of V_2O_5 caused a price rise to around $40/kg, which in turn caused a rise in the price of ferrovanadium.

Due to its high thermal coefficient of resistance, vanadium (V) oxide finds use as a detector material in bolometers and microbolometer arrays for thermal imaging.

Possible new uses include the preparation of bismuth vanadate ceramics for use in solid oxide fuel cells.

Biological Activity: Despite being highly toxic in humans, vanadium occurs in nature for example in enzymes such as vanabins. Vanadate (VO_4^{3-}), formed when V_2O_5 is dissolved in water at alkaline pH, appears to inhibit enzymes that process phosphate (PO_4^{3-}). However the exact mode of action remains elusive .

Precautions: Vanadium (V) oxide is toxic.

Element Class

The Elemental group includes metals and intermetallic elements (gold, silver, copper), semi-metals and non-metals (antimony, bismuth, graphite, sulphur). This group also includes natural alloys, such as electrum (a natural alloy of gold and silver), phosphides, silicides, nitrides and carbides (which are usually only found naturally in a few rare meteorites).

Electrum

Electrum is a naturally occurring alloy of gold and silver, with trace amounts of copper and other metals. Colour ranges from pale to bright yellow, depending on the proportions of gold and silver. Gold content of naturally-occurring electrum in modern Western Anatolia ranges from 70 per cent to 90 per cent in contrast to the 45-55 per cent of electrum used in ancient Lydian coinage of the same geographical area.

Electrum was used as early as the third millennium BC in Old Kingdom Egypt, sometimes as an exterior coating to the pyramidions atop ancient Egyptian pyramids and obelisks.

Electrum was also used in the making of ancient drinking vessels and coins.

Composition: Electrum consists primarily of gold and silver but is sometimes found with traces of copper and other metals. As a result, electrum is usually a good conductor of electricity.

Analysis of the electrum composition in ancient Greek coinage dating from 600 BC showed that the gold composition was 55.5 per cent in archaic Phocaea. In the early classical period, the gold composition of electrum ranged from 46 per cent in Phokaia to 43 per cent in Mytilene. In later coinage from these areas, dating to 326 BC, the gold composition averaged 40 per cent to 41 per cent.

Appearance: The colour of electrum is pale yellow or yellowish-white and the name is a Latinised form of the Greek word (*elektron*) mentioned in the *Odyssey* meaning a metallic substance consisting of gold alloyed with silver. The same word was also used for the substance amber, probably because of the pale yellow colour of certain varieties, and it is from the electrostatic properties of amber that the modern English words "electron" and "electricity" derive.

Electrum was often referred to as white gold in ancient times but could be more accurately described as "pale gold". The modern use of the term white gold usually concerns gold, silver and palladium alloys.

History

Electrum is mentioned in an expedition sent by Pharaoh Sahure of the Fifth dynasty of Egypt.

Electrum is believed to have been used in coins circa 600 BC in Lydia under the reign of Alyattes II.

Electrum was much better for coinage than gold, mostly because it was harder and more durable, but also because techniques for refining gold were not widespread at the time. The discrepancy between gold content of electrum from modern Western Anatolia (70-90 per cent) and ancient Lydian coinage (45-55 per cent) suggests that the Lydians had already solved the refining technology for silver and were adding Ag to the local native electrum some decades before introducing the pure silver coins cited below.

In Lydia, 14.1 grams of electrum was made into one *stater* (meaning "Standard"). A stater was about one month's pay for a soldier. To complement the stater, fractions were made: the *trite* (third), the *hekte* (sixth), and so forth, including 1/24 of a stater, and even down to 1/48th and 1/96th of a stater. The 1/96 stater was only about 0.14 to 0.15 grams.

Because of the variety of electrum's composition, it was rather difficult to determine the exact worth of each coin. Widespread trading was somewhat hampered by this, as a foreign merchant would offer rather poor rates on local electrum coin.

These difficulties were eliminated in 570 BC when pure silver coins were introduced. However, electrum currency remained fairly popular until approximately 350 BC. The simplest reasoning for this would be that, because of the gold content, one 14.1 gram stater would be worth as much as ten 14.1 gram silver pieces.

In fiction: In the fictional *Star Wars* universe, Jedi Master Mace Windu wielded a lightsaber that had a hilt made of electrum (only members of the Jedi Council were permitted to make such a saber). This was cosmetic only. The Sith Lord Darth Sidious also constructed a lightsaber that had a hilt made of electrum.

A naturally-occurring golden alloy named electrum featured in the *Transformers* episode, *The Golden Lagoon,* but featured the additional unusual property of rendering Transformers immune to damage when coated in it. However, it proved to be vulnerable to intense levels of heat.

Before 3rd Edition, an electrum coin (electrum piece) was worth five silver pieces (half a gold piece) in the Dungeons & Dragons role-playing game. Its removal was the final part of the decimalisation of D&D currency.

The Terry Pratchett book *Pyramids* features an Electrum-plated Pyramidion that was used in a failed attempt to "flare off" the time stored in a magical pyramid.

In the *Rex Mundi* comic, electrum is a rare metal valued by sorcerers for its ability to focus magical energies.

One of Jaden Yuki's monsters from anime series *Yu-Gi-Oh! GX* is named Elemental Hero Electrum (Although it's Japanese name is Elixirer), named so because of its gold and silver armour.

In alchemy and (occult) magic, Electrum is an alloy of equal parts of all the planetary metals.

In *Final Fantasy XII,* the abnormally high-level and rare enemies called Entites will leave behind Electrum if they are vanquished by the element counter to their own elemental alignment.

In Garth Nix's series, *The Keys to the Kingdom,* electrum is often used to make transfer plates.

Organic Class

The organic mineral class includes biogenic substances in which geological processes have been a part of the genesis or origin of the existing compound. Minerals of the organic class include various oxalates, mellitates, citrates, cyanates, acetates, formates, hydrocarbons and other miscellaneous species. Examples include whewellite, moolooite, mellite, fichtelite, carpathite and abelsonite.

Whewellite

Whewellite is a mineral, hydrated calcium oxalate, formula $CaC_2O_4{\cdot}H_2O$. Because of its organic content it is thought to have an indirect biological origin and this is supported by it being found in coal and sedimentary nodules. However, it has also been found in hydrothermal deposits where a biological source appears improbable. For this reason it can be classed as true mineral.

Whewellite, or at least crystalline calcium oxalate does also arise from natural sources. Small crystals or flakes are sometimes found forming on the surface of some cacti. Also some particularly unpleasant kidney stones are deposited as calcium oxalate.

Whewellite is named after William Whewell (1794-1866), an English mineralogist.

Moolooite

Moolooite is a rare blue-green mineral with the formula $Cu^{++}(C_2O_4){\cdot}n(H_2O)$ $(n<1)$ (making it a hydrated copper oxalate). It was discovered by RM Clarke and IR Williams in Bunbury Well, Mooloo Downs station, Murchison, Western Australia in 1986. It has an orthorhombic crystalline structure, and is formed by the interaction of bat guano with weathering copper sulphides.

Mellite

Mellite, also called honeystone, is an unusual mineral being also an organic chemical. Chemically identified as an aluminium salt of mellitic acid, that is, aluminium benzene hexacarboxylate hydrate, with the chemical formula $Al_2C_6(COO)_6{\cdot}16H_2O$.

It is a translucent honey-coloured crystal which can be polished and faceted to form striking gemstones. It crystallises in the tetragonal system and occurs both in good crystals and as formless masses. It is soft with a Mohs hardness of 2 to 2.5 and has a low specific gravity of 1.6. It was discovered originally in 1789 at Artern in Thuringia in Germany it has subsequently also been found in Russia, Austria, the Czech Republic, and Hungary. It is found associated with lignite and is assumed to be formed from plant material. Although chemically different it is in many ways analogous to amber which is formed the same way.

FICHTELITE

Fichtelite is a rare white mineral found in fossilised wood from Bavaria. It crystallises in the monoclinic crystal system. It is a cyclic hydrocarbon dimethyl-isopropyl-perhydropenanthrene, $C_{19}H_{34}$. It is very soft with a Mohs hardness of 1, the same as talc. Its specific gravity is very low at 1.032, just slightly more dense than water. It was first described in 1841 and named for the location, Fichtelgebirge, Bavaria, Germany.

ABELSONITE

Abelsonite is a mineral of nickel, carbon, hydrogen, and nitrogen with formula: $Ni[C_{32}H_{36}N_4]$. It forms purple to reddish brown triclinic crystals. The crystals are soft, with a Mohs scale rating is 2 to 3. It has a low specific gravity of 1.45.

It was first described in 1975 for specimens in oil shale from the Green River Formation, eastern Uintah County, Utah. It is named after Philip Hauge Abelson, an American physicist.

HOW THE MINERALS ARE FORMED AND CHANGED

Observations of the weathering zones of ore deposits prove that the rate of the sulphide oxidation depends on many factors among which the following should be mentioned:

- Mineral composition of ores (generally *marcasite* oxidises easier than pynte and polymineral aggregates easier than monomineral);
- Size of grains (fine-grained aggregates oxidise easier than coarse-grained);
- Composition and structure of the wall-rocks;
- Climatic and hydrogeological conditions (fide e.g. Smirnov 1956).

Many of these factors take also part in laboratory oxidation of iron sulphides. Coarse-grained, pure pyrite aggregates are the best preserved ones in mineral collections and still remain fresh, while many of fine-crystalline aggregates of marcasite already after few years undergo a complete decay.

However, the observations of specimens of the same mineral with very similar development, stored under the same conditions, revealing that one of them is almost unaltered whereas the other is covered by thick efflorescences of secondary sulphates, suggest that an additional factor plays a substantial role in such a case. It seems that the differences in the rate of weathering of FeS2 specimens are connected with the mechanism and local conditions of oxidation.

The pathways of FeS2 oxidation may be controlled either chemically or organically by microorganisms, especially the bacteria Thiobacillus ferrooxidans (fide Moses et al. 1987; Taylor and Wheeler 1984, 1994). Under favourable conditions the bacterial oxidation rate is accelerated up to few orders of magnitude relative to the abiotic rate.

One can notice in the collection under study that the oxidation process runs most intensively in those specimens which contain kaolinite clay impurities or in which iron sulphides crystallised on the lignite base. Kaolinite and lignite play here twofold role. On the one hand they absorb water, an indispensable agent in each oxidation mechanism. Under laboratory conditions, usually with low humidity, the supply of water to the oxidation environment is of fundamental importance (Borek 1994). The possibility of keeping strong acidic environment in the porous matter is of equal importance. Very low pH values are necessary for activity of the oxidation bacteria.

These bacteria can intensively develop only in such places where sulphuric acid resulting from the oxidation of sulphides is preserved. If the medium is not porous (i.g. coarse-crystalline sulphides without admixture of other substances), the progress of oxidation is significantly slower.

The oxidation of FeS2 encrusting siderite matrix runs more slowly because siderite acts as a neutralising agent for sulphuric acid. Therefore, the specimens of marcasite covering the siderite matrix have been preserved without sign of oxidation, whereas similar specimens of marcasite on the lignite matrix were subject to the complete decomposition.

It seems that marcasite specimens on the carbonate matrix (e.g. marl, limestone, dolomite, siderite) are the only proper specimens for mineral collections. A chance of keeping marcasite specimens on the lignite or clay matrix for a longer period is rather small.

The observed differences in the rate of sulphide alteration can be explained by different conditions of the growth of oxidation bacteria. The iron sulphides altered very insignificantly in the specimens when the conditions were not favourable for developing of bacteria.

But in any place where the originated acid products stabilised the best conditions of the bacteria development, oxidation was significantly accelerated. As a result, pyrite or marcasite altered quickly into powdery assemblages of sulphate minerals. Simultaneously, the originating sulphuric acid destroys labels and cardboard boxes in which such specimens are kept.

The crystallisation of the sulphate paragenesis results from evaporation of small portion of solutions concentrating the products of FeS2 oxidation and clay alteration in acidic environment. The oxidation occurs not only on the surface but mainly inside the specimens where from the solutions are drawn up through capillary fractures and then concentrate on the surface.

The kind of the sulphate minerals is a function of the solution composition and local conditions, e.g. humidity, acidity and redox potential. One can notice the evolution of the mineral composition of the sulphate efflorescences during storage. It is similar to that observed under natural condition and lead to the crystallisation of more stable phases containing ferric iron. An inconspicuous presence of rozenite, commonly found in similar parageneses, is the characteristic feature of the specimens studied.

Moreover, the presence of aluminium sulphates, e.g. halotrichite and alunogen, deserves attention, with halotrichite appearing in the earlier stage of weathering and then being replaced by alunogen.

4

Mineral Family

Titanium Minerals

Titanium is a chemical element; in the periodic table it has the symbol Ti and atomic number 22. It is a light, strong, lustrous, corrosion-resistant (including resistance to sea water and chlorine) transition metal with a white-silvery-metallic colour. Titanium can be alloyed with other elements such as iron, aluminium, vanadium, molybdenum and others, to produce strong lightweight alloys for aerospace (jet engines, missiles, and spacecraft), military, industrial process (chemicals and petro-chemicals, desalination plants, pulp and paper), automotive, agri-food, medical (prostheses), sporting goods, and other applications. Titanium was discovered in England by William Gregor in 1791 and named by Martin Heinrich Klaproth for the Titans of Greek mythology.

The element occurs within a number of mineral deposits, principally rutile and ilmenite, which are widely distributed in the Earth's crust and lithosphere, and found in almost all living things, rocks, water bodies and soils. The metal is extracted from its principal mineral ores via the Kroll process.

Its most common compound, titanium dioxide, is used in the manufacture of white pigments. Other compounds include titanium tetrachloride (used in smoke screens/skywriting and as a catalyst) and titanium trichloride (used as a catalyst in the production of polypropylene).

The two most useful properties of the metal form are corrosion resistance, and the highest strength-to-weight ratio of any metal. In its unalloyed condition, titanium is as strong as steel, but 45 per cent lighter. There are two allotropic forms and five naturally occurring isotopes of this element; ^{46}Ti through ^{50}Ti with ^{48}Ti being the most abundant (73.8 per cent). Titanium's properties are chemically and physically similar to zirconium.

A metallic element, titanium is recognised for its high strength-to-weight ratio. It is a light, strong metal with low density that, when pure, is quite ductile (especially in an oxygen-free environment), lustrous, and metallic-white in colour. The relatively high melting point (over 1,649 °C or 3,000 °F) makes it useful as a refractory metal.

Commercial (99.2 per cent pure) grades of titanium have ultimate tensile strengths of about 63,000 psi, equal to that of steels alloys, but are 45 per cent lighter. Titanium is 60 per cent heavier than aluminium, but more than twice as strong as the most commonly used 6061-T6 aluminium alloy. Certain titanium alloys (e.g., Beta C) achieve tensile strengths of over 200,000 psi (1.4 GPa). However, titanium loses strength when heated above 430 °C (800 °F).

It is fairly hard (although by no means as hard as some grades of heat-treated steel) and can be tricky to machine due to the fact that it will gall if sharp tools and proper cooling methods are not used. Like those made from steel, titanium structures have a fatigue limit which guarantees longevity in some applications.

The metal is a dimorphic allotrope with the hexagonal alpha form changing into the body-centred cubic (lattice) beta form at 882 °C (1,619 °F). The heat capacity of the alpha form increases dramatically as it is heated to this transition temperature but then falls and remains fairly constant for the beta form regardless of temperature.

CHEMICALS

The most noted chemical property of titanium is its excellent resistance to corrosion; it is almost as resistant as platinum, capable of withstanding attack by acids, moist chlorine gas, and by common salt solutions. Pure titanium is not soluble in water but is soluble in concentrated acids.

This metal forms a passive and protective oxide coating (leading to increased corrosion-resistance) when exposed to elevated temperatures in air, but at room temperatures it resists tarnishing. When it first forms, this protective layer is only 1 to 2 nanometres thick but continues to slowly grow; reaching a thickness of 25 nanometres in four years.

Titanium burns when heated in air 610 °C (1,130 °F) or higher, forming titanium dioxide. It is also one of the few elements that burns in pure nitrogen gas (it burns at 800 °C or 1,472 °F and forms titanium nitride, which causes embrittlement). Titanium is resistant to dilute sulphuric and hydrochloric acid, along with chlorine gas, chloride solutions, and most organic acids. It is paramagnetic (weakly attracted to magnets) and has fairly low electrical and thermal conductivity.

Experiments have shown that natural titanium becomes radioactive after it is bombarded with deuterons, emitting mainly positrons and hard gamma rays. When it is red hot the metal combines with oxygen, and when it reaches 550 °C (1,022 °F) it combines with chlorine. It also reacts with the other halogens and absorbs hydrogen.

The processing of titanium metal occurs in 4 major steps: reduction of titanium ore into "sponge", a porous form; melting of sponge, or sponge plus a master alloy to form an ingot;

primary fabrication, whereby an ingot is converted into general mill products such as billet, bar, plate, sheet, strip and tube; and secondary fabrication of finished shapes from mill products.

Because the metal reacts with air at high temperatures it cannot be produced by reduction of its dioxide. Titanium metal is therefore produced commercially by the Kroll process, a complex and expensive batch process. The relatively high market value of titanium is mainly due to its processing, which sacrifices another expensive metal, magnesium. In the Kroll process, the oxide is first converted to chloride through carbochlorination, whereby chlorine gas is passed over red-hot rutile or ilmenite in the presence of carbon to make $TiCl_4$. This is condensed and purified by fractional distillation and then reduced with 800 °C molten magnesium in an argon atmosphere.

A more recently developed method, the FFC Cambridge process, may eventually replace the Kroll process. This method uses titanium dioxide powder (which is a refined form of rutile) as feedstock to make the end product which is either a powder or sponge. If mixed oxide powders are used, the product is an alloy manufactured at a much lower cost than the conventional multi-step melting process. The FFC Cambridge Process may render titanium a less rare and expensive material for the aerospace industry and the luxury goods market, and could be seen in many products currently manufactured using aluminium and specialist grades of steel.

Common titanium alloys are made by reduction. For example; cuprotitanium (rutile with copper added is reduced), ferrocarbon titanium (ilmenite reduced with coke in an electric furnace), and manganotitanium (rutile with manganese or manganese oxides) are reduced.

$$2TiFeO_3 + 7Cl_2 + 6C\ (900\ °C) \rightarrow 2TiCl_4 + 2FeCl_3 + 6CO$$

$$TiCl_4 + 2Mg\ (1100\ °C) \rightarrow 2MgCl_2 + Ti$$

About 50 grades of titanium and titanium alloys are designated and currently used, although only a couple of dozen are readily available commercially. The ASTM International recognises 31 Grades of titanium metal and alloys, of which Grades 1 through 4 are commercially pure (unalloyed). These four are distinguished by their varying degrees of tensile strength, as a function of oxygen content, with Grade 1 being the most ductile (lowest tensile strength with an oxygen content of 0.18 per cent), and Grade 4 the least (highest tensile strength with an oxygen content of 0.40 per cent). The remaining grades are alloys, each designed for specific purposes, be it ductility, strength, hardness, electrical resistivity, creep resistance, resistance to corrosion from specific media, or a combination thereof.

The grades covered by ASTM and other alloys are also produced to meet Aerospace and Military specifications (SAE-AMS, MIL-T) ISO standards, and country-specific specifications, as well as proprietary end-user specifications for aerospace, military, medical and industrial applications.

In terms of fabrication, all welding of titanium must be done in an inert atmosphere of argon or helium in order to shield it from contamination with atmospheric gases such as oxygen, nitrogen or hydrogen. Contamination will cause a variety of conditions, such as embrittlement, which will reduce the integrity of the assembly welds and lead to joint failure. Commercially pure flat product (sheet, plate) can be formed readily, but processing must take into account the fact that the metal has a 'memory' and tends to spring back. This is especially true of certain high-strength alloys. The metal can be machined using the same equipment and via the same processes as stainless steel.

Applications

Titanium is used in steel as an alloying element (ferro-titanium) to reduce grain size and as a deoxidiser, and in stainless steel to reduce carbon content. Titanium is often alloyed with aluminium (to refine grain size), vanadium, copper (to harden), iron, manganese, molybdenum, and with other metals. Applications for titanium mill products (sheet, plate, bar, wire, forgings, castings) can be found in industrial, aerospace, recreational and emerging markets.

Minerals containing the chemical element titanium are called Tatanium Minerals.

AENIGMATITE

Aenigmatite is a sodium, iron, titanium inosilicate mineral with formula: $Na_2Fe^{2+}{}_5TiSi_6O_{20}$. It forms brown to black triclinic lamellar crystals. It has Mohs hardness of 5.5 to 6 and specific gravity of 3.74 to 3.85. Aenigmatite forms a solid-solution series with wilkinsonite, $Na_2Fe^{2+}{}_4Fe^{3+}{}_2Si_6O_{20}$.

Aenigmatite is primarily found in peralkaline volcanic rocks, pegmatites, and grantes as well as silica-poor intrusive rocks. It was first described in 1865 for an occurrence in the Ilimaussaq intrusive complex of west Greenland. Its name comes from the Greek word for "riddle".

It was also reported from the Kaidun meteorite, possibly a Mars meteorite, landed on March 1980 in South Yemen. Other notable studied occurrences include: Narssarssukand elsewhere in Greenland in the Khibiny and Lovozero massifs, Kola Peninsula, Russia, from Pantelleria, Italy. In the USA, from Granite Mountain, near Little Rock, Pulaski Co., Arkansas, and Santa Rosa, Sonoma Co., California, in Australia, from Warrumbungle volcano, Nandewar volcano, and the Mt. Warning complex, New South Wales; and the Peak Range Province, Queensland, from Logan Point quarry, Dunedin volcano, New Zealand.

AESCHYNITE-(Y)

Aeschynite-(Y) (or Aeschinite-(Y), Aeschynite-(Yt), Blomstrandine, Priorite) $(Y,Ca,Fe,Th)(Ti,Nb)_2(O,OH)_6$ is a mineral of yttrium, calcium, iron, thorium, titanium, niobium, oxygen, and hydrogen. Its name comes from the Greek word for "*shame*". Its Mohs scale rating is 5 to 6.

ANATASE

Anatase is one of the three mineral forms of titanium dioxide (the other two being brookite and rutile). It is always found as small, isolated and sharply developed crystals, and like rutile, a more commonly occurring modification of titanium dioxide, it crystallises in the tetragonal system; but, although the degree of symmetry is the same for both, there is no relation between the interfacial angles of the two minerals, except, of course, in the prism-zone of 45° and 90°. The common pyramid of anatase, parallel to the faces of which there are perfect cleavages, has an angle over the polar edge of 82°9', the corresponding angle of rutile being 56°52½'. It was on account of this steeper pyramid of anatase that the mineral was named, by RJ Hauy in 1801, from the Greek anatasis, "extension," the vertical axis of the crystals being longer than in rutile. There are also important differences between the physical characters of anatase and rutile; the former is not quite so hard (H=5½-6) or dense (specific gravity 3.9); it is optically negative, rutile being positive; and its lustre is even more strongly adamantine or metallic-adamantine than that of rutile.

Two types or habits of anatase crystals may be distinguished. The commoner occurs as simple acute double pyramids with an indigo-blue to black colour and steely lustre. Crystals of this kind are abundant at Le Bourg-d'Oisans in Dauphiné, where they are associated with rock-crystal, feldspar, and axinite in crevices in granite and mica-schist. Similar crystals, but of microscopic size, are widely distributed in sedimentary rocks, such as sandstones, clays, and slates, from which they may be separated by washing away the lighter constituents of the powdered rock.

Crystals of the second type have numerous pyramidal faces developed, and they are usually flatter or sometimes prismatic in habit; the colour is honey-yellow to brown. Such crystals closely resemble xenotime in appearance and, indeed, were for a long time supposed to belong to this species, the special name wiserine being applied to them. They occur attached to the walls of crevices in the gneisses of the Alps, the Binnenthal near Brig in canton Valais, Switzerland, being a well-known locality.

When strongly heated, anatase is converted into rutile, changing in specific gravity to 4.1; naturally occurring pseudomorphs of rutile after anatase are also known. Crystals of anatase have and continue to be artificially prepared in laboratories by introducing the moisture-sensitive titanium tetrachloride, $TiCl_4$, to water at very cold temperatures (the process is very exothermic) to produce TiO_2 and HCl gas. Such synthetic forms of anatase are currently under scrutiny in the field of semi-conductors and photovoltaic materials.

Another name commonly in use for this mineral is octahedrite, a name which, indeed, is earlier than anatase, and given because of the common (acute) octahedral habit of the crystals. Other names, now obsolete, are oisanite and dauphinite, from the well-known French locality.

Armalcolite

Armalcolite is a mineral that was discovered at Tranquility Base on the Moon by the Apollo 11 crew in 1969. It was named for Armstrong, Aldrin and Collins, the three Apollo 11 astronauts. Examples of this mineral were later discovered in various locations on Earth as well, such as Smoky Butte, Garfield County, Montana, USA in lamproite dikes and plugs; Disko Island, (Greenland) in native iron; Pripyat Swell (around the Pripyat River, Ukraine) in trachybasalt and in South African kimberlites. Armalcolite has a chemical formula of $(Mg,Fe^{++})Ti_2O_5$. It has a Mohs hardness of 6 and a specific gravity of 4. It forms grey sub to anhedral masses with a crystal structure that is orthorhombic prismatic. Armalcolite is a part of the pseudobrookite group which consists of minerals of the general formula X_2YO_5. X and Y are usually Fe (2+ and 3+), Mg, Al, and Ti. End members are armalcolite ($MgFeTi_2O_5$), pseudobrookite (Fe_2TiO_5), and ferropseudobrookite ($FeTi_2O_5$).

Brookite

Brookite is a mineral consisting of titanium oxide, TiO_2, and hence identical with rutile and anatase in composition, but crystallising in the orthorhombic system.

It was named for Henry James Brooke, English mineralogist, 1771 - 1857.

Euxenite

Euxenite or euxenite-Y is a brownish black mineral with a metallic lustre, found in Norway. It contains calcium, niobium, tantalum, cerium, titanium, yttrium, and typically uranium and thorium, with some other metals. The chemical formula is: $(Y,Ca,Ce,U,Th)(Nb,Ta,Ti)_2O_6$. It occurs in pegmatites and black sands. Commonly partially amorphous due to radiation damage.

Haggertyite

Haggertyite is a rare barium, iron, magnesium, titanate mineral: $Ba(Fe^{2+}{}_6Ti_5Mg)O_{19}$ first described in 1996 from the Crater of Diamonds near Murfreesboro in Pike County, Arkansas. The microscopic metallic mineral crystallises in the hexagonal system and forms tiny hexagonal plates associated with richterite and serpentinitised olivine of mafic xenoliths in the lamproite host rock. It is an iron (II) rich member of the magnetoplumbite group. It is a light grey mineral with Mohs hardness of 5.

It was named for Stephen E. Haggerty (b. 1938), of the University of Massachusetts.

Hibonite

Hibonite ($(Ca,Ce)(Al,Ti,Mg)_{12}O_{19}$) is a brownish black mineral with a hardness of 7.5-8.0 and a hexagonal crystal structure. It is rare, but is found in high-grade metamorphic rocks on Madagascar. Some pre-solar grains in primitive meteorites consists of Hibonite.

Keilhauite

Keilhauite (also known as yttrotitanite) is a mineral of a brownish black colour, related to titanite in form. It consists chiefly of silicon dioxide, titanium dioxide, calcium oxide, and yttrium oxide.

It differs from titanite only in that calcium is substituted by up to 12 per cent $(Y,Ce)_2O_3$.

Nenadkevichite

Nenadkevichite is a rare silicate mineral containing niobium with formula: $(Na,Ca)(Nb,Ti)Si_2O_7 \cdot 2H_2O$. It forms brown to yellow to rose coloured orthorhombic dipyramidal crystals with a dull to earthy lustre. It has a Mohs hardness of 5 and a specific gravity of 2.86.

It was first reported in 1955 from a nepheline syenite pegmatite in the Kola Peninsula. In addition it has been reported from Mt. Saint-Hilaire, Canada; the Ilimaussaq complex, Greenland; Windhoek District, Namibia; and Zheltye Vody, Ukraine. It was named after Konstantin A. Nenadkevich, Russian mineralogist and geochemist.

Enstatite

The pyroxene silicate minerals enstatite ($MgSiO_3$) and ferrosilite ($FeSiO_3$) form a complete solid solution series and are common rock-forming minerals found in igneous and metamorphic rocks and meteorites. The intermediate composition, $(Mg,Fe)SiO_3$, is sometimes known as hypersthene although this name has been formally abandoned and replaced by orthopyroxene, and when determined, giving relative proportions of enstatite (En) and ferrosilite (Fs) (e.g., $En_{80}Fs_{20}$). Weathered enstatite with a small amount of iron takes on a submetallic lustre and a bronze-like colour. This material is termed bronzite although it is more correctly called altered enstatite. Most natural crystals are orthorhombic (space group Pbcn) although three polymorphs are known. The high temperature, low pressure polymorphs are protoenstatite and protoferrosilite (also orthorhombic, space group Pbcn) while the low temperature forms, clinoenstatite and clinoferrosilite, are monoclinic (space group $P2_1/c$).

Bronzite and hypersthene were known long before enstatite, which was first described by G. A. Kenngott in 1855.

An emerald green variety of enstatite is called chrome-enstatite and is cut as a gemstone. The green colour is caused by traces of chromium, hence the varietal name. In addition, bronzite is also sometimes used as a gemstone.

Identification

Enstatite and the other orthorhombic pyroxenes are distinguished from those of the monoclinic series by their optical characteristics, such as straight extinction, much weaker double refraction and stronger pleochroism. They also have a prismatic cleavage that is perfect in two directions at 90 degrees. Enstatite is white, gray, greenish or brown in colour; its hardness is 5 - 6, and its specific gravity is 3.2 - 3.3.

Occurrence

Isolated crystals are rare, but orthopyroxene is an essential constituent of various types of igneous rocks and metamorphic rocks. Magnesian orthopyroxene occurs in plutonic rocks such as gabbro (norite) and diorite. It may form small idiomorphic phenocrysts and also groundmass grains in volcanic rocks such as basalt, andesite, and dacite.

Enstatite, close to $En_{90}Fs_{10}$ in composition, is an essential mineral in typical peridotite and pyroxenite of the Earth's mantle. Xenoliths of peridotite are common in kimberlite and in some basalt. Measurements of the calcium, aluminium, and chromium contents of enstatite in these xenoliths have been crucial in reconstructing the depths from which the xenoliths were plucked by the ascending magmas.

Orthopyroxene is an important constituent of some

metamorphic rocks such as granulite. Orthopyroxene near pure enstatite in composition occurs in some metamorphosed serpentines. Large crystals, a foot in length and mostly altered to steatite, were found in 1874 in the apatite veins traversing mica-schist and hornblende-schist at the apatite mine of Kjorrestad, near Brevig in southern Norway.

Enstatite is a common mineral in meteorites. Crystals have been found in stony and iron meteorites, including one that fell at Breitenhach in the Erzgebirge, Bohemia. In some meteorites, together with olivine it forms the bulk of the material; it can occur in small spherical masses, or chondrules, with an internal radiated structure.

POLYCRASE

Polycrase or polycrase-(Y) is a black or brown metallic complex uranium yttrium oxide mineral with formula: $(Y,Ca,Ce,U,Th)(Ti,Nb,Ta)_2O_6$. It crystallises in the orthorhombic system with typically radial prismatic crystal form. It has a Mohs hardness of 5 to 6 and a specific gravity of 5. It is radioactive due to its uranium content (around 6 per cent). It occurs in granitic pegmatites.

It was first described in 1870 at Rasvag, Hidra (Hittero) Island, near Flekkefjord, Norway. It is found in Sweden, Norway, and the United States.

ZIRCOPHYLLITE

Zircophyllite is a complex mineral, formula $(K,Na)_3(Mn,Fe)_7(Zr,Ti,Nb)_2Si_8O_{24}(OH,F)_7$. It crystallises in the triclinic - pinacoidal crystal class as dark brown to black micaceous plates. It has perfect 001 cleavage, a Mohs hardness of 4 to 4.5 and a specific gravity of 3.34. Its indices of refraction are $n\alpha = 1.708$ $n\beta = 1.738$ $n\gamma = 1.747$ and it has a 2V optical angle of 62°.

It occurs with natrolite in alkali pegmatites. It was discovered in 1972 in the Korgeredabinsh massif, Tuva, Russia and is named for its zirconium content and its relationship to astrophyllite. It is also known from the Mt. Saint-Hilaire intrusive complex of Quebec, Canada.

Zirkelite

Zirkelite is an oxide mineral with formula: (Ca,Th,Ce) $Zr(Ti,Nb)_2O_7$. It occurs as well-formed fine sized isometric crystals. It is a black mineral with a hardness of 5.5 and a specific gravity of 4.7.

Zirkelite was first discovered in Brazil in 1895. It was named for Ferdinand Zirkel (1838-1912), German petrographer.

Initial discovery was from the Jacupiranga carbonatite, Sao Paulo, Brazil. It is also found in Canada, Kazakhstan, Norway, Russia, South Africa, the United Kingdom, and the USA.

Clay Minerals

Clay minerals form an important group of the phyllosilicate or sheet silicate family of minerals, which are distinguished by layered structures composed of polymeric sheets of SiO_4 tetrahedra linked to sheets of (Al, Mg, Fe)(O,OH)6 octahedra. The geochemical importance of clay minerals stems from their ubiquity in soils and sediments, high specific surface area, and ion exchange capacities. Clay minerals tends to dominate the surface chemistry of soils and sediments. Furthermore, these properties give rise to a wide range of industrial applications throughout the history of mankind.

The use of clay for mainly clay figures, pottery and ceramics was already known by primitive people about 25000 years ago (Shaikh and Wik, 1986). Today clay is an important material with a large variety of applications in ceramics, oil drilling, liners for waste disposal, and the metal and paper industry. Clay is furthermore used as adsorbent, decolouration agents, ion exchanger, and molecular sieve catalyst (Fowden et al., 1984; Murray, 1991). Despite their importance, the clay minerals form a difficult group of minerals to study due to their small size, variable structural composition, and relative slow kinetics of formation and alteration.

CLASSIFICATION

Smectites are phyllosilicates or layer silicates having a layer lattice structure in which two-dimensional oxoanions are separated by layers of hydrated cations. The oxygen atoms define upper and lower sheets enclosing tetrahedral sites, and a central sheet having the brucite or gibbsite structure enclosing octahedral sites. Smectites having two tetrahedral sheets around the central octahedral sheet relation are known as 2:1 phyllosilicates. Kaolinite, on the other hand, has one tetrahedral and one octahedral sheet (1:1).

A further designation can be made based on the type and location of the cations in the oxygen framework. In one unit cell composed of twenty oxygen atoms and four hydroxyl groups, there are eight tetrahedral and six octahedral sites. A smectite is dioctahedral if two-thirds of the octahedral sites are occupied by trivalent cations, and trioctahedral if all octahedral sites are filled with bivalent cations. Micas, though no smectites, have identical 2:1 phyllosilicate oxygen frameworks.

In talc and pyrophyllite all tetrahedral sites are filled with Si_4+ and the octahedral sites either completely with Mg2+ or for 2/3 with Al3+, respectively. The electrically neutral sheets are bonded together by relatively weak dipolar and van der Waals forces (Giese, 1975). In contrast, smectite layers have a positive charge deficiency resulting from the isomorphous substitutions, viz., (i) Si4+ by Al3+ at tetrahedral sites, and (ii) Al3+ by Mg2+, or (iii) Mg2+ by Li+ (or a vacancy) at octahedral sites. The charge deficiency is balanced by hydrated interlayer cations, such as, Na+, K+, or Ca2+.

The charge deficiency of smectites is intermediate between that of micas and that of pyrophyllite and talc. Differences the charge of the layers, the origin of the charge deficiency, and interlayer cations results in different physical and chemical properties, such as, thermal stability and swelling behaviour. The layer charge of an octahedrally substituted smectite (e.g. montmorillonite) is distributed over the complete oxygen framework, whereas tetrahedral substitution (as e.g. in beidellite) leads to a more localised charge distribution and the last smectites tend to a higher three-dimensional order (Brindley, 1980; Suquet et al., 1975).

The Abrasives

An abrasive is a material, often a mineral, that is used to shape or finish through rubbing which leads to part of the workpiece being worn away. While finishing a material often means polishing it to gain a smooth, reflective surface it can also involve roughening the abraded substance as in satin, matt or beaded finishes.

Abrasives are extremely commonplace and are used very extensively in a wide variety of industrial, domestic, and technological applications. This gives rise to a large variation in the physical and chemical composition of abrasives as well as the shape of the abrasive. Common uses for abrasives include grinding, polishing, buffing, honing, cutting, drilling, sharpening and sanding. (For simplicity, "mineral" in this article will be used loosely to refer to both minerals and mineral-like substances whether man-made or not.)

Files act by abrasion but are not classed as abrasives as they are a shaped bar of metal. However, diamond files are a form of coated abrasive (as they are metal rods coated with diamond powder). Abrasives give rise to a form of wound called an abrasion or even an excoriation. Abrasions may arise following strong contract with surfaces made things such as concrete, stone, wood, carpet and roads, though these surfaces are not intended for use as abrasives.

Mechanics of Abrasion

Abrasives generally rely upon a difference in hardness between the abrasive and the material being worked upon, the abrasive being the harder of the two substances. However, this is not necessary as any two solid materials that repeatedly rub against each other will tend to wear each other away (such as softer shoe soles wearing away wooden or stone steps over decades or centuries or glaciers abrading stone valleys).

Typically, materials used as abrasives are either hard minerals (rated at 7 or above on Mohs scale of mineral hardness) or are synthetic stones, some of which may be chemically and physically identical to naturally occurring minerals but which cannot be called minerals as they did not arise naturally. (While useful for comparative purposes, the Mohs scale is of limited value to materials engineers as it is an arbitrary, ordinal, irregular scale.) Diamond, a common abrasive, for instance occurs both naturally and is industrially produced, as is corundum which occurs naturally but which is nowadays more commonly manufactured from bauxite. However, even softer minerals like calcium carbonate are used as abrasives, such as "polishing agents" in toothpaste.

These minerals are either crushed or are already of a sufficiently small size (anywhere from macroscopic grains as large as about 2 mm to microscopic grains about 0.001 mm in diameter) to permit their use as an abrasive. These grains, commonly called grit, have rough edges, often terminating in points which will decrease the surface area in contact and increase the localised contact pressure. The abrasive and the material to be worked are brought into contact while in relative motion to each other. Force applied through the grains causes fragments of the worked material to break away while simultaneously smoothing the abrasive grain and/or causing the grain to work loose from the rest of the abrasive.

Some factors which will affect how quickly a substance is abraded include:

- Difference in hardness between the two substances: a much harder abrasive will cut faster and deeper
- Grain size (grit size): larger grains will cut faster as they also cut deeper

- Adhesion between grains, between grains and backing, between grains and matrix: determines how quickly grains are lost from the abrasive and how soon fresh grains, if present, are exposed
- *Contact force:* more force will cause faster abrasion
- *Loading:* worn abrasive and cast off work material tends to fill spaces between abrasive grains so reducing cutting efficiency while increasing friction
- *Use of lubricant/coolant/metalworking fluid:* Can carry away swarf (preventing loading), transport heat (which may affect the physical properties of the workpiece or the abrasive), decrease friction (with the substrate or matrix), suspend worn work material and abrasives allowing for a finer finish, conduct stress to the workpiece.

Abrasive Minerals

Abrasives may be classified as either natural or synthetic. When discussing sharpening stones, natural stones have long been considered superior but advances in material technology are seeing this distinction become less distinct. Many synthetic abrasives are effectively identical to a natural mineral, differing only in that the synthetic mineral has been manufactured rather than been mined. Impurities in the natural mineral may make it less effective.

Some naturally occurring abrasives are:

- Calcite (calcium carbonate)
- Emery (impure corundum)
- Diamond dust (synthetic diamonds are used extensively)
- Novaculite
- Pumice dust
- Rouge
- Sand

Some abrasive minerals (such as zirconia alumina) occur naturally but are sufficiently rare or sufficiently more difficult/ costly to obtain such that a synthetic stone is used industrially. These and other artificial abrasives include:

- Borazon (Cubic Boron Nitride or CBN)
- Ceramic
- Corundum (alumina or aluminium oxide)
- Dry ice
- Glass powder
- Silicon carbide (carborundum)
- Zirconia alumina

Manufactured Abrasives

Abrasives are shaped for various purposes. Natural abrasives are often sold as dressed stones, usually in the from of a rectangular block. Both natural and synthetic abrasives are commonly available in a wide variety of shapes, often coming as bonded or coated abrasives, including blocks, belts, discs, wheels, sheets, rods and loose grains.

Bonded Abrasives

A bonded abrasive is composed of an abrasive material contained within a matrix, although very fine aluminium oxide abrasive may comprise sentried material. This matrix is called a binder and is often a clay, a resin, a glass or a rubber. This mixture of binder and abrasive is typically shaped into blocks, sticks, or wheels. The most usual abrasive used is aluminium oxide. Also common are silicon carbide, tungsten carbide and garnet. Artificial sharpening stones are often a bonded abrasive and are readily available as a two sided block, each side being a different grade of grit.

Grinding wheels are cylinders that are rotated at high speed. While once worked with a foot pedal or hand crank, the introduction of electric motors has made it necessary to construct the wheel to withstand greater radial stress to prevent the wheel flying apart as it spins. Similar issues arise with cutting wheels which are often structurally reinforced with impregnated fibres. High relative speed between abrasive and workpiece often makes necessary the use of a lubricant of some kind.

Traditionally they were called coolants as they were used to prevent frictional heat build up which could damage the workpiece (such as ruining the temper of a blade). Some research suggests that the heat transport property of a lubricant is less important when dealing with metals as the metal will quickly conduct heat from the work surface. More important are their effects upon lessening tensile stresses while increasing some compressive stresses and reducing "thermal and mechanical stresses during chip formation".

Various shapes are also used as heads on rotary tools used in precision work, such as scale modelling.

Bonded abrasives need to be dressed after they are used. Dressing is cleaning the waste material (swarf and loose abrasive) from the surface and exposing fresh grit. Depending upon the abrasive and how it was used, dressing may involve the abrasive being simply placed under running water and brushed with a stiff brush for a soft stone or the abrasive being ground against another abrasive, such as aluminium oxide used to dress a grinding wheel.

Truing is restoring the abrasive to its original surface shape. Wheels and stones tend to wear unevenly, leaving the cutting surface no longer flat (said to be "dished out" if it is meant to be a flat stone) or no longer the same diameter across the cutting face. This will lead to uneven abrasion and other difficulties.

Coated Abrasive

Coated abrasives are made of abrasive grains adhered to the surface of flexible or semi-flexible backings such as paper, cloth, vulcanised fibre, plastic films. They are used for DIY as well as industrial applications.

Coated abrasives grit sizes range from very coarse (~2 mm) to ultrafine (submicron). Most grit sizes are standardised at the world level by ISO standards (ISO 6344). Coated abrasives are used by hand or on machines, applications are extremely varied, at home and in the industry for:

- Fine finishes: hand sanding or polishing of wooden cabinets, car bodies primers, metals.
- High precision operations: optical lens polishing, magnetic head superfinishing.
- Medium stock removal with portable or fixed machines: floor sanding, disc grinding of metal, endless belts grinding of wood, metal, glass.
- Heavy stock removal with wide (up to three metres) belts on high power machines (up to 500 hp) to grind stainless steel sheets or coils or to calibrate plywood or chipboards panels.

Sandpaper and emery cloth are coated abrasives for hand use, usually non- precision. These two terms are used by general public in place of "coated abrasives".

Other Abrasives and their Uses

Sand, glass beads, metal pellets and dry ice may all be used for a process called sandblasting (or similar, such as the use of glass beads which is "bead blasting"). Dry ice will sublimate meaning that there is no residual abrasive left afterwards.

Cutting compound used on automotive paint is an example of an abrasive suspended in a liquid, paste or wax, as are some polishing liquids for silverware and optical media. The liquid, paste or wax acts as a binding agent that keeps the abrasive attached to the cloth which is used to as a backing to move the abrasive across the workpiece. On cars in particular, wax may serve as both a protective agent by preventing exposure of the paint of metal to air and also act as an optical filler to make scratches less noticeable. Toothpaste contains calcium carbonate or silica as a "polishing agent" to remove plaque and other matter from teeth as the hardness of calcium carbonate is less than that of tooth enamel but more than that of the contaminating agent.

Very fine rouge powder was commonly used for grinding glass, being somewhat replaced by modern ceramics, and is still used in jewellery making for a highly reflective finish.

Cleaning products may also contain abrasives suspended in a paste or cream. They are chosen to be reasonably safe on some linoleum, tile, metal or stone surfaces. However, many laminate surfaces and ceramic topped stoves are easily damaged by these abrasive compounds. Even ceramic/pottery tableware or cookware can damage these surfaces, particularly the bottom of the tableware which is often unglazed in part or in whole and acts as simply another bonded abrasive.

Metal pots and stoves are often scoured with abrasive cleaners, typically in the form of the aforementioned cream or paste or of steel wool. Human skin is also subjected to abrasion in the form of exfoliation. Abrasives for this can be much softer and more exotic than for other purposes and may include things like almond and oatmeal. Dermabrasion and microdermabrasion are now rather commonplace cosmetic procedures which use mineral abrasives.

Scratched compact discs and DVDs may sometimes be repaired through buffing with a very fine compound, the principle being that a multitude of small scratches will be more optically transparent than a single large scratch. However, this does take some skill and will eventually cause the protective coating of the disc to be entirely eroded (especially if the original scratch is deep), after which the data surface will be destroyed if abrasion continues.

Choice of Abrasive

The shape, size and nature of the workpiece and the desired finish will influence the choice of the abrasive used. A bonded abrasive grind wheel may be used to commercially sharpen a knife (producing a hollow grind), but an individual may then sharpen the same knife with a natural sharpening stone or a even flexible coated abrasive (like a sandpaper) stuck to a soft, non-slip surface to make achieving a convex grind easier. Similarly, a brass mirror may be cut with a bonded abrasive, have its surface flattened with a coated abrasive to achieve a basic shape, and then have finer grades of abrasive successively applied culminating in a wax paste impregnated with rouge to leave a sort of "grainless finish" called, in this case, a "mirror finish".

Also, different shapes of adhesive may make it harder to abrade certain areas of the workpiece. Health hazards can arise from any dust produced (which may be ameliorated through the use of a lubricant) which could lead to silicosis (when the abrasive or workpiece is a silicate) and the choice of any lubricant. Besides water, oils are the most common lubricants. These may present inhalation hazards, contact hazards and, as friction necessarily produces heat, flammable material hazards. An abrasive which is too hard or too coarse can remove too much material or leave undesired scratch marks. Besides being unsightly, scratching can have other, more serious effects. Excessive abrasion or the presence of scratches may:

- Diminish or destroy usefulness (as in the case of scratched optics and compact discs or a dull knife);
- Trap dirt, water, or other material;
- Increase surface area (permitting greater chemical reactivity such as increased rusting which is also affected by matter caught in scratches);
- Erode or penetrate a coating (such as a paint or a chemical or wear resistant coating);
- Overlay quickly cause an object to wear away (such as a blade or a gemstone);
- Increase friction (as in jewelled bearings and pistons).

A finer or softer abrasive will tend to leave much finer scratch marks which may even be invisible to the naked eye (a "grainless finish"); a softer abrasive may not even significantly abrade a certain object. A softer or finer abrasive will take longer to cut as tends to cut less deeply than a coarser, harder material. Also, the softer abrasive may become less effective more quickly as the abrasive is itself abraded.

This allows fine abrasives to be used in the polishing of metal and lenses where the series of increasingly fine scratches tends to take on a much more shiny or reflective appearance or greater transparency. Very fine abrasives may be used to coat the strop for a cut-throat razors, however, the purpose of stropping is not to abrade material but to straighten the burr on an edge. The final stage of sharpening Japanese swords called polishing and may be a form of superfinishing.

Different chemical or structural modifications may be made to alter the cutting properties of the abrasive.

Other very important considerations are price and availability. Diamond, for a long time considered the hardest substance in existence, is actually softer than fullerite and even harder aggregated diamond nanorods, both of which have been synthesised in laboratories but no commercial process has yet been developed. Diamond itself is expensive due to scarcity in nature and the cost of synthesising it. Bauxite is a very common ore which, along with corundum's reasonably high hardness, contributes to corundum's status as a common, inexpensive abrasive.

Thought must be given to the desired task about using an appropriately hard abrasive. At one end, using an excessively hard abrasive wastes money by wearing it down when a cheaper, less hard abrasive would suffice. At the other end, if too soft, abrasion does not take place in a timely fashion, effectively wasting the abrasive as well as any accruing costs associated with loss of time.

Other Instances of Abrasion: Aside from the

aforementioned uses of shaping and finishing, abrasives may also be used to prepare surfaces for application of some sort of paint of adhesive. An excessively smooth surface may prevent paint and adhesives from adhering as strongly as an irregular surface could allow. Inflatable tyre repair kits (which, on bicycles particularly, are actually patches for the inner tube rather than the tyre) require use of an abrasive so that the self-vulcanising cement will stick strongly.

Inadvertently, people who use knives on glass or metal cutting boards are abrading their knife blades. The pressure at the knife edge can easily create microscopic (or even macroscopic) cuts in the board. This cut is a ready source of abrasive material as well as a channel full of this abrasive through which the edge slides. For this reason—without regard for the health benefits—wooden boards are much more desirable. A similar occurrence arises with glass-cutters. Glass-cutters are have circular blades that are designed to roll not slide. They should never retrace an already effected cut.

Undesired abrasion may result from the presence of carbon in internal combustion engines. While smaller particles are readily transported by the lubrication system, larger carbon particles may abrade components with close tolerances. The carbon arises from the excessive heating of engine oil or from incomplete combustion. This soot may contain fullerenes which are noted for their extreme hardness—and small size and limited quantity which would tend to limit their effect.

5

Formation of Minerals

Formation and Growth of Crystal Phases

A crystalline solid may form in different ways: (a) by the crystallisation of liquids (melts or solutions); (b) by the deposition of crystals on walls of openings from the gaseous products of sublimation, and (c) by the recrystallisation of solids (e.g., colloids). The crystalline formations occurring in nature are overwhelmingly the products of crystallisation of silicate melts and aqueous solutions. This includes the enormous masses of igneous crystalline rocks, the vast majority of commercial mineral deposits, and the crystalline precipitates of saline basins.

Theoretically the crystallisation of a cooling melt should begin at a definite temperature corresponding to the melting point of a given substance. In the same way a solution should crystallise at the point when the solvent is saturated with the solute. Actually, however, the crystallisation of liquid phases begins at certain *supercooling* or *supersaturation.*

The degree of supercooling or supersaturation required

for the crystallisation to begin depends on the chemical composition and, to some extent, on pressure. Pressure variations play a much more important role when the crystals are formed out of condensing vapours.

The process of the growth of crystals in both a supercooled melt and in a supersaturated solution is exactly the same. The initiation of crystals may be *compelled,* if some fragments and specks of a solid which act as primers, due to their crystallochemical properties, are already present in the liquid; it may also *be spontaneous,* in the absence of any primers in a supersaturated solution or supercooled melt.

In the case of spontaneous crystallisation at different points of the melt or solution there appear so-called *crystallisation centres* or crystal seeds. At the beginning of crystallisation (e.g., in a single-component liquid), regular crystals will grow unimpeded around the crystallisation centres until they impinge on one another, thus obstructing the further development of crystallographic forms. Further crystallisation takes the form of struggle for the remaining space, and the end product is an aggregate of irregularly-shaped crystal grains. Sometimes the grains show a crystalline zonality, as evidence of their gradual growth.

There is a direct relationship between the degree of supercooling or supersaturation and the number of crystallisation centres which spontaneously develop when solidification begins: the more supercooled or supersaturated is the liquid, the *greater* will be the *number* of crystallisation centres in a given volume per unit of time, and hence the smaller will be the crystal grains on final solidification.

If crystallisation begins at relatively low supersaturation of the solution $k\pm$ 35, the number of developing crystal nuclei will be comparatively small and the resulting aggregates will be relatively coarsely crystalline. If crystallisation begins at greater super-saturation (or supercooling), e.g., at k_2, the natural result will be finely-crystalline aggregates. Upon crystallisation of an extremely supersaturated solution, cryptocrystalline substances or colloids will be formed.

According to experimental data, initiation of crystals in a liquid depends on a variety of factors, such as the chemical character of the material, the presence of impurities which may accelerate or delay the capillary channels. The following experiment is a good illustration. If a hygroscopic cotton thread is immersed in a saturated solution of table salt, the part of the thread exposed to air will develop a crystalline aggregate fed by the saturated solution rising by capillary action. In the air, the solution becomes strongly supersaturated through evaporation, which results in the crystallisation of the solute on the cotton thread.

If the solvent evaporates slowly, well-developed crystals may form in this way. The student may remember seeing ice needles growing from damp soil after a clear frosty night. Similar needles grow in the air from slowly drying damp powders of salts, which are readily soluble in water, e.g., calcium chloride. This may be the mechanism of the formation of long acicular crystals of many minerals found in cavities.

Crystals may also form in a gaseous medium without being fed by a liquid solution, when a substance transmutes directly from a vapour into a solid at certain temperatures (below the melting point) and pressures. One example is the formation of snow flakes in the air in the form of stellate crystals. Another example are the various minerals deriving from volcanic sublimation.

Frequently, crystals and crystal grains contain tiny inclusions of foreign matter (solid, liquid and gaseous). These inclusions for the most part make the crystals appear dim or opaque. They can be readily seen upon microscopic examination in polished sections. Only when a mineral is uniformly opaque it is rather difficult to discern the liquid or gaseous inclusions.

The study of the spatial distribution of these foreign substances shows that they were apparently mechanically entrapped by the rapidly growing crystals. Inside the crystals they are often oriented in crystallographic directions. Examples of this are the inclusions of volcanic glass (solidified magma drops) in crystalline-banded plagioclases, mother liquors of the salts of K, Na, Ga, etc., gas bubbles which are usually mixed with a liquid in dim crystals of quartz, calcite, topaz, and other minerals.

It is curious that gas-liquid inclusions, when heated to a definite temperature, usually become a homogeneous liquid (the gas dissolving in the liquid) and, when cooled, the gas bubble separates again. This is important because in this way one may determine the approximate temperature at which the mineral crystallises and captures the minute drops of the solution.

Some of the gas-liquid inclusions may even contain a third, solid phase (e.g., tiny NaCl crystals). On heating these crystals are the first to dissolve in the liquid, followed by the gas bubble. In addition to primary gas-liquid inclusions, the crystals may have later, secondary inclusions, which are associated with "healed" cracks in the crystals. It will be noted that in the secondary gas-liquid inclusions the gas bubbles in the drops of liquid will disappear upon heating earlier than in primary inclusions. On further heating the crystal mass will decrepitate in either case (so great is the pressure of the expanding liquid inclusions).

If a solution impregnating a rock enters into an exchange decomposition reaction with it, this, as a rule, results in neo-crystallisation involving either all of the rock or some of its constituents. This process known as *replacement* or *metasomatism,* is exemplified in the replacement of calcite by gypsum upon reaction with water containing sulphuric acid:

$$CaCO_3 + H_2O + H_2SO_4 = CaSO_4 \bullet 2H_2O + CO_2$$

calcite gypsum or in the replacement of sphalerite by covellite upon reaction with copper-sulphate solution:

$$ZnS + CuSO_4 = CuS + ZnSO_4$$

sphalerite solution covellite solution

In the case of selective metasomatism (when only some of the minerals in a rock are replaced) the resulting mineral, which retains not only the external shape but sometimes also the peculiar internal structure of the original mineral, is called a *metasome*. In many cases when the constituent replaced is a crystal there develops a *pseudomorph*, i.e., a crystal form which is quite foreign to the new mineral. When organic remains, such as wood, are replaced by colloids like opal or iron sulphides, their peculiar structure often remains intact.

Quite common in nature are well-formed crystals developed through metasomatism in solid environments (rocks). Such formations, or *metacrystal Sj* occur in a limited number of minerals. An example are the well-formed cubic crystals of pyrite in schists, marbles and other rocks. Metacrystals often contain remnants of unreplaced minerals of the enclosing rock. Not infrequently they occur along hardly detectable hair cracks in rocks, which definitely points to their later origin.

Recrystallisation and other transformations of minerals in *solid* media occur as a result of material changes in the physico-chemical equilibrium, in particular in conditions of so-called regional metamorphism.

Solution and Decomposition of Minerals: It has already been stated that many minerals undergo diverse transformations because of changes in the environment, sometimes dissolving, completely or decomposing into insoluble products of chemical reactions.

The beginning of *solution* is easily detected in individual crystals by the following characteristic features:

(1) Whereas during the growth of a crystal it is the apexes and edges that tend to grow faster, during solution they also dissolve faster, and therefore the crystal becomes somewhat rounded;

(2) Whereas during the growth of a crystal the slowly growing faces of a crystal are the most stable, on solution of a crystal fragment or a sphere made out of a crystal

the first to appear will be the faces that dissolve most rapidly;

(3) The slowly growing faces usually have brilliant and smooth surfaces; on solution the slowly dissolving faces often appear lustreless;

(4) At the initial moments of solution, the faces often develop very shallow polyhedral solution pits or etch figures.

Decomposition of minerals in nature, either complete or incomplete, is mainly associated with oxidation and reduction processes. This is especially true of minerals containing elements which under natural conditions are capable of forming several ions of different valencies (e.g., Fe^{2+}, Fe^{3+}, Mn^{2+}, Mn^{3+}, Mn^{4+}, S^{2-}, etc.).

When a mineral, which originally contained lower valency cations, finds itself in oxidising conditions (e.g., the weathering zone of rocks and ores) these cations will naturally tend to pass into ions of higher valency. During this process the cations diminish in size with the subsequent disintegration of the crystal structure. Thus, in an oxygen and water-rich environment, the divalent iron cation in FeS (pyrrhotite) readily becomes trivalent with the formation of poorly soluble iron hydroxides, whereas the divalent sulphur anion oxidises into a hexavalent cation with the formation of a complex $[SO_4]^2$ anion, which, on reacting with hydrogen ions, finally yields sulphuric acid, which goes into solution. In this way, a new substance with completely different properties is formed in place of pyrrhotite. In the same way, a carbonate of divalent manganese ($MnCO_3$) readily forms hydroxides of tetravalent manganese.

If iron and manganese hydroxides, in the course of geological processes, find themselves deep in the earth's crust where a reducing environment prevails, the higher valencies of the cations of elements often change into lower valencies, which is accompanied by the dehydration of the compounds.

New minerals are formed under these new conditions: haematite (Fe_2O_3) or magnetite ($Fe—Fe_2—O_4$), braunite (Mn-$Mn—O_3$) hausmannite ($Mn—Mn_2—O_4$), etc.

Generations of Minerals: The generations of a mineral, as the term implies, are its differently aged formations in a particular mineral association, differing also in size, appearance, and specific features of chemical composition. Observations of the occurrence of minerals, especially in ore deposits, often point to the existence of several generations of the same mineral developing within a single stage of mineral genesis.

In some instances younger generations of finer crystals overgrow older and larger crystals of the same mineral. In other instances, one observes older coarse-grained generations and younger fine-grained crystalline formations along thin cracks of other minerals, or, as micro-inclusions discernible only under the microscope, etc. Detailed microscopic examination of ores shows that almost every ore-forming mineral comprises several generations, which indicates the complexity of the processes of ore-genesis. This complexity must not be ignored in studying the paragenesis of minerals.

Mineral Aggregates: The crystallisation and solidification of a solution or melt gives rise to a mixture of intergrown grains called a mineral aggregate. Aggregates may be monomineral consisting of crystal grains of a single mineral (e.g., marble or magnetite) and polymineral, consisting of several minerals of different properties and composition (e.g., granite or copper-zinc sulphide ore).

Mineral aggregates occur in a wide variety of forms and structures, some of them being so characteristic that they are designated by special terms. The main morphological feature of a mineral aggregate is the degree of its crystallinity. With regard to crystallinity, minerals fall into two clearly distinct groups:

(1) Phanerocrystalline aggregates, and

(2) Cryptocrystalline and colloidal masses.

The principal types of mineral aggregates are the following:

1. *Granular aggregates* composed of crystal grains sometimes occurring together with well-developed crystals of some minerals. This type is the most widespread in the earth's crust. Examples: holocrystalline igneous rocks, numerous sulphides and other ores.

With regard to grain size, aggregates are classed as (1) coarse-grained, with grains over 5 mm in diameter; (2) medium-grained, with grains 1 to 5 mm in diameter, readily distinguished by the naked eye; and (3) fine-grained, with grains less than 1 mm in diameter.

The structure of cryptocrystalline aggregates can be established only under the microscope in thin polished sections.

The shape of the grains comprising the aggregate is reflected in the morphological features of the aggregate as a whole. If an aggregate consists of roughly equidimensional grains, it is simply called *granular*.

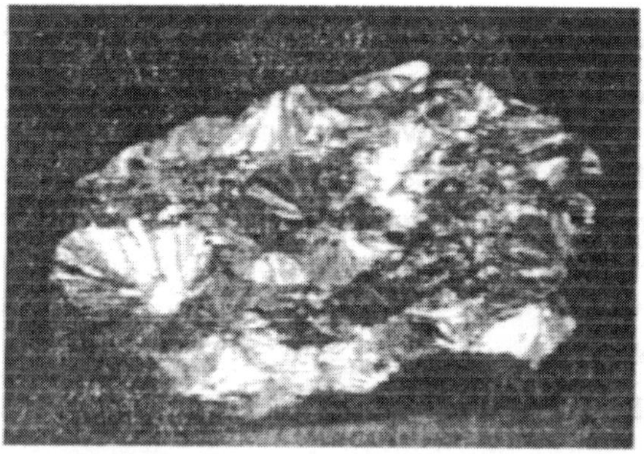

Radially-fibrous aggregates of pyrophyllite.

If the grains are lamellar in habit, the aggregate is called *foliated* or *scaly*, depending on the size of the grains. Aggregates with grains elongated in one direction, sometimes radiating or diverging from a common centre are called *columnar, acicular,* or *fibrous*. Also widespread are aggregates composed of minerals of different forms, such as mica schists with isometric garnet crystals, granular masses of quartz with columnar tourmaline crystals, etc.

An aggregate may be compact or friable. An example of the latter is crystalline formations at the bottom of drying salt lakes.

Druses are intergrowths of well-developed crystals on the walls of cavities. An example are the common druses of quartz crystals. Druses are interesting not only crystallographically but also because they often slow the sequence of segregation of different minerals crystallising from the last portions of a solution.

The very fact that druses contain well-developed crystals indicates that they have formed in free space, that is, in some primary cavities, hollow cracks, fractured rocks, etc. These cavities may range in size from tiny pores to big caves whose walls are covered with large crystals of transparent quartz and other minerals.

Mineral formations, in which the crystal individuals are in close contact and elongated parallel to each other, are referred to as *comb* or *drusy* aggregates. Evidently, the close contact between the crystals forming on the walls of cavities allowed them to develop only in a direction perpendicular to the walls. This group also includes *crystalline crusts*, composed of fine, closely intergrown crystals.

It has been demonstrated that in these cases the numerous original crystal nuclei develop in random directions. However, in the process of growth, fewer and fewer crystal individuals will survive in the competition for available space, and it will be those growing perpendicular to the surface of initial crystal that will have the greatest chance of survival ("geometrical selection"). If the surface is concave, the result will be converging radiated aggregates. If the surface is convex, diverging acicular or columnar crystalline masses will form.

Secretions are formed when an irregularly-shaped, but generally spherical cavity is filled with crystalline or colloidal material. A peculiar feature of many secretions is a concentrical deposition of mineral matter from the walls of a cavity towards its centre. The layers thus deposited frequently vary in colour and sometimes in composition.

Small openings or cavities are usually completely filled with mineral matter. Sometimes the central part is filled with radially-fibrous aggregates of some mineral, e.g., zeolite. Large cavities often have a hollow centre with druses or sinter formations lining its walls.

Smaller secretions (up to 10 mm in diameter) are termed *amygdules,* while bigger ones are called *geodes. Concretions* are globular or irregular spherical formations and nodules which develop in unconsolidated sedimentary rocks, mostly in clays, sands, and in the earthy products of rock weathering.

They vary widely in size, from a few millimetres to tens of centimetres and sometimes even metres in diameter. As they expand and coalesce, the concretions form large bodies of complex configuration.

Often, though not as a rule, concretions develop around foreign bodies, usually organic remains. In polished sections the sandy concretions of phosphorite and marcasite show a banded arrangement of sand particles, which corresponds to the banded structure of the rock as a whole. This shows that the concretions were formed, at least partially, later than the enclosing rock. In all probability they first appeared as colloidal clots or gels, which subsequently crystallised. When fractured across the centre, they often display a radially-fibrous texture. Occasionally the mineral mass displays indistinct concentric banded structure. Thus, the origin of concretions differs materially from that of secretions. In contrast to the latter, concretions grow out radially from a centre. The minerals found most often as concretions are phosphorite, pyrite, marcasite, sometimes siderite, barite, etc.

Oolites form in almost the same way as concretions. They also are spherical formations, but somewhat smaller in size (from tenths of a millimetre to 5-10 mm), developing in an aqueous environment around suspended foreign matter, such as sand particles, fragments of organic remains, and even gas bubbles. A characteristic feature of oolitic formations is their distinct, rather regular concentric banded, occasionally conchoidal, structure. Formations analogous in shape, but lacking concentric banding are so-called pseudolites.

Recent calcareous oolites are formed from suspended matter in running water, and when they attain a certain size, they settle to the bottom. Sedimentary rocks composed of cemented pea-sized oolites are called pisolites.

Sinter forms of minerals originate from colloids or gels. Like crystal druses, they are found in cavities. When slowly migrating colloidal solutions enter a cavity, they coat its walls. Gradually losing water (dispersion medium), which evaporates into the empty space inside the cavities, the solutions thicken and under the action of gravity form *stalactitic*, *reniform* 46, *botryoidal*, and other shapes which finally solidify. Falling colloidal drops may deposit tapered *stalagmites* on the floors of cavities.

Such formations range widely in size, from microscopic to huge, pillar-like stalactites and stalagmites of aragonite and calcite ($CaCO_3$) in large caves. Most diverse minerals may appear in the forms of sinters iron hydroxides (limonite, goethite), manganese hydroxides (psilomelane group), opal, malachite, gypsum, aragonite, calcite, sulphides of different metals, etc. Examination of sinter masses (in cross-sections) often reveals a concentric zonal structure.

This is due to alternation of zones of the same minerals which are differently coloured or possess different physical properties (malachite, limonite, etc.) or much more rarely because the zones consist of minerals of different composition (limonite, chalcedony, and malachite; limonite and native copper, etc.). The differences in the mineral composition of the concentric bands reflect changes in the composition of the solutions, which produced the sinter forms.

It has been stated earlier that colloidal formations or gels crystallise rather easily. Since the process takes a different course with different minerals, the resulting crystalline aggregates will also differ in structure. For instance, limonite sinter formations, while fully retaining their external shape, turn into an aggregate of diverging fine fibres of goethite or hydrohaematite perpendicular to the surface of the individual concentric layers. Other minerals crystallise into coarser radiated fibrous aggregates, which nevertheless show traces of the original concentric layers on fracture. In this process the smooth surface of sinter formations becomes studded with crystal facets (marcasite).

Earthy masses, as the name implies, are soft mealy formations in which no crystallinity is detectable even with a hand lens. Usually they are found as crusts or accumulations which are most often the products of chemical weathering. Depending on colour, such masses are sometimes called *sooty* (black) or *ochreous* (yellow or brown accumulations and crusts).

Examples are variously coloured nickel hydrosilicates which occur as earthy mineral formations; sooty manganese hydroxides; ochreous iron hydroxides, and other residual products of weathering.

Coatings and selvages which sometimes occur as thin films on crystal surfaces may consist of most diverse substances. These include films of brown iron hydroxides on rock crystals; the green or blue selvages on the enclosing rocks of copper deposits, etc.

Efflorescences are friable coatings and crusts or sporadic moss like and fluffy formations of readily soluble hydrosulphates or other salts appearing periodically on the surface of ores, rocks, dry soils and in cracks. They usually disappear during the rainy seasons to reappear with dry weather. Another type of efflorescences are the very common dendritic growths of manganese hydroxides on the surface of rocks along thin cracks.

Liesegang Rings and Spirals: This term denotes regularly alternating banded formations which result from the intermittent deposition of compounds in the course of diffusion in gels, very similar to the concentric rings and spirals obtained by R. LiesQgang. Essentially, his experiment consisted of the following: microscopic $Ag_2Cr_2O_7$ crystals were formed around an $AgNO_3$ drop on gelatin impregnated with $K_2Cr_2O_7$. The crystals at first moved with the solution, but later as they grew larger, were retained in the gelatin pores and deposited at regular intervals to form concentric rings.

Structures of this kind are sometimes found in agates and jaspers. Very similar formations develop in finely porous rocks, in the course of weathering processes. Examples are the periodic rings, bands and hyperbolas stained by brown iron hydroxides found in limestones, sandstones, and other rocks. In this case the staining matter apparently is deposited from sols in the form of a gel when the dispersed phase or the electrolyte reaches the critical concentration point. If this process goes parallel with the leaching of a rock, the end result will be concentric conchoidal formations in which bands of compact gel alternate with earthy bands. Ores with alternating bands of sphalerite and ankerite or magnetite and calcite belong to this type of formations.

Sometimes, in place of rings and spirals, dendritic formations are found, such as iron and manganese hydroxide inclusions in opal 8 - 8010 113 ("moss" agates). They can also be readily reproduced artificially in a gelatinous medium.

Paragenesis of Minerals: This term, which was understood as "the simultaneous occurrence of minerals", was introduced into geology in 1849 by Breithaupt. Long before Breithaupt, however, V. M. Severgin (1798) in Russia had used the term "contiguity of minerals", which is another way of saying the same thing. In 1923, Vernadsky coined the term "mineral association" for the simultaneous occurrence of several minerals in the same mineral body, in distinction from the term "paragenesis" to which he attached a different meaning.

He believed that this idea had its beginnings in the observations of early miners who looked for accessory minerals as guides to ores and gemstones. "Though not aware of that, they were in fact studying mineral associations." Thus, it has long been known that galena (PbS) which is often associated with silver, as a rule, occurs together with sphalerite (ZnS). Similarly, in many deposits gold is associated with quartz, and cinnabar (HgS) with antimonite (Sb_2S_3). The vast amount of empirical evidence which has already been collected in this field is a great aid to geological prospecting and exploration.

By "paragenesis" Vernadsky meant "all the mineral associations known for a given mineral or chemical element", and he suggested that this should embrace the mineral bodies in which the given mineral is found, its mineral associations and generations. Unfortunately, Vernadsky did not cite concrete examples to illustrate his concept. Elsewhere in his work, however, Vernadsky stressed the need for "studying the dependences governing the combination of minerals or their paragenesis". Viewed in this context the problem acquires special interest to mineralogy.

The vast amount of mineralogical data collected in Soviet times in the process of detailed study of various rocks and ores has greatly furthered research in these directions. Soviet scientists have established that definite paragenetic associations, that is, *groups of minerals formed together* in a given ijimeral body develop in the process of mineral genesis, depending on the physico-chemical conditions and the interaction of solutions with the environment, at each stage of the process. Characteristically, each such group reflects the particular conditions under which the minerals had formed. The following example will illustrate this point.

Often one and the same ore specimen may have among its constituents *two or more groups of minerals which differ in the time of their formation and origin.* Thus limonite (iron hydroxide) and malachite (copper carbonate) often occur together with semi-decomposed copper and iron sulphides (pyrite—FeS_2 and chalcopyrite—$CuFeS_2$). Geological data always show, however, that the sulphides formed earlier under one set of conditions, while the iron hydroxides and copper carbonate formed later under entirely different conditions (in conditions of weathering). The two groups are spatially associated only because they had a common source of the chemical elements entering into their composition (iron and copper). Hence, in this mineral association we have *two groups of minerals differing in the conditions of their formation.*

This approach to mineral associations enables conclusions to be drawn about the *dependences, important both theoretically and practically, governing the alteration of paragenetic associations of minerals with time* and bearing evidence to changes in the physico-chemical conditions in the course of mineral genesis. Soviet scientists developed a geometrical method for the analysis of different mineral associations occurring in nature which makes it possible to reveal many details and facts which generally elude the student when other methods are used.

Knowledge of typical paragenetic associations is very important in mineralogy. Apart from its value as an aid to the identification of minerals occurring together, it is an important guide to economic minerals. Thus, if the easily identifiable minerals pyrrhotite (FeS) and chalcopyrite ($CuFeS_2$) are found in a basic magnesial igneous rock, one must look for a third, nickeliferrous mineral, pentlandite, which is not as readily indentifiable but is very important economically.

In nature, minerals occur in most diverse paragenetic associations. This depends not only on the initial composition of the crystallising solutions or the enclosing rocks which react with them, but also on such factors as temperature, pressure, depth of the formation or transformation of the minerals, etc. Some minerals develop at very specific environmental factors, while others may form under various conditions.

Thus gypsum ($CaSO_4$-$2H_2O$) under some conditions occurs in association with chlorides and sulphates in salt beds formed as a result of the drying of salt lakes and lagoons. In other instances it is the product of chemical weathering (in areas with scarce precipitation) in association with iron hydroxides and the argillaceous products of disintegration. Mining experience shows that it disappears quite rapidly with depth. Furthermore, gypsum has been found as crystals in fissures, in decomposed and discoloured lava and igneous rocks in the vicinity of solfa-taras where it might have derived from calcareous minerals through the action of hot waters containing sulphuric acid, etc.

The diversity of paragenetic mineral associations is often further complicated by the fact that superimposed on a group of minerals formed in one process may be mineral associations formed through other processes, the later minerals not infrequently deriving partly from those of the earlier group. Hence the importance, in such cases, of taking separate account of each *group of contemporaneous minerals,* since each process of mineral formation, and consequently each group, has *its own* dependences governing its mineral associations. The very fact that one process is superimposed on another is evidence of a change in the physico-chemical equilibrium of the mineral phases. Owing to this, at least some of the earlier formed minerals become unstable under the new conditions and are therefore either altered or replaced by other minerals.

As far as typical associations of chemical elements in separate minerals are concerned, this problem, in the case of crystalline bodies, is solved entirely on the basis of the laws of crystal chemistry. The combination of the different elements within a crystalline body depends upon the chemical properties, i.e., on the structure of atoms or ions, as well as on their sizes and properties. The replacement of some atoms or groups of atoms by others likewise depends on their sizes. In ionic compounds the condition is added that the sum total of positive and negative valencies must be balanced.

Typomorphic Features of Minerals: It has long since been noticed that some minerals have crystal forms typical of a given deposit or rock. Thus calcite ($CaCO_3$) which occurs as crystals usually in cavities occurs in different deposits as scalenohedral, obtusely or acutely rhombohedral, platy, fine rod-like, etc. Attempts were made to establish a connection between these different crystal forms and the temperature of their formation. In the end, however, it was found that the morphology and sizes of the crystals depended not so much on temperature, as on the concentration of the constituents in solutions, the presence of certain dissolved impurities, and the degree of supersaturation.

At the present time not only the crystal forms, but any typical features of minerals such as colour, admixtures of chemical elements, types of twinning, etc., are regarded as "typomorphic" features. There is no question that the characteristics of a particular deposit are related to the specific features of the solutions, from which its minerals have crystallised, as well as to temperature, pressure and other factors of mineral formation.

It has been noted that cassiterite (SnO_2) from high-temperature or pegmatite formations (in distinction from that of hydrothermal deposits) often Contains as impurities such metals as columbium, tantalum, iron, etc. Therefore, if found as rounded grains in unconsolidated deposits of gullies and river valleys, the specific features of its composition give a rough indication of the type of deposit it has come from.

It has been also established that large quartz crystals from Alpine clefts, quartz from other deposits, are characterised by certain features of crystal form, specific paragenetic association and the presence of relatively large monocrystal blocks inside crystals which are usually twinned according to the Dauphine law.

Native gold is characteristically richest in silver (isomorphously admixed) when found in deposits, which have formed in the upper part of the earth's crust, i.e., at relatively low temperatures and pressures. Argentiferous gold (electrum) differs from ordinary native gold in physical properties as well (lower specific gravity and lighter yellow colour). Sulphur compounds of silver, such as argentite (Ag_2S) and proustite (Ag_3AsS_3), etc., are often paragenetically associated with electrum.

The vast majority of ore deposits of a complex origin show a great variety of typomorphic features, which necessitates very detailed research to establish their true dependences as the basis for positive conclusions. Be that as it may, paragenetic associations remain important criteria in the study of mineral deposits.

Geological Processes of Mineral Formation

To establish the genesis (origin) of a given complex of mineral it is essential not only to know the way they have formed, but also the geological processes that occur in the earth's crust and give rise to rocks and ores of most diverse composition. Since these problems are thoroughly covered by special courses of petrography and deposits of economic minerals, the question of mineral genesis will be dealt with here only in general terms.

All mineral masses which are formed as a result of some geologic process fall into two basic genetic groups, depending on the source of energy to which they owe their formation:

(1) *Endogenous* (born from within) formed in the course of processes which depend on the internal thermal energy of the earth; the minerals of this group are the products of magmatic activity (in the broad sense of the term); rocks and economic mineral deposits result from the crystallisation of the magma itself and from different segregations from the latter; mineral forming processes occur at different depths and at different, though usually high, temperatures;

(2) *Exogenous* (born from without) depending for their formation on the energy of the sun; the source of the mineral matter are outcropping and disintegrating rocks and ores of different origin; mineral formation processes

occur in the uppermost part of the earth's crust at low temperatures and pressures close to atmospheric, in the conditions of interaction between the physical and chemical agents of the atmosphere, hydrosphere, and biosphere.

After their formation, both the endogenous and exogenous mineral masses undergo transformations under conditions of altered environment (metamorphism). Particularly profound transformations, both in composition and structure, occur in the course of *regional metamorphism* when the rock masses and the associated deposits are moved by tectonic dislocations from their zone of formation to the deeper zones of the earth's crust.

Endogenous Processes of Mineral Formation

Modern knowledge of endogenous processes is based on the concepts of the activity of magmatic hearths or chambers situated in the lower parts of the earth's crust. These processes occur at great depths, and cannot be observed directly. Only in the vicinity of active volcanoes can we obtain certain data indicating the nature of the processes occurring in the interior of the earth. Moreover, knowledge of the composition, structural features, occurrence, and interrelation of various igneous rocks and mineral deposits spatially associated with them enables some idea to be formed (in agreement with the physico-chemical laws) of the dependences governing the endogenous processes of mineral formation.

According to this concept magmas are silicate melts of complex composition containing also some volatile constituents.

When for some reason or other large masses of magma fail to reach the surface, they cool slowly in the upper part of the earth's crust under high external pressure. At the same time magma differentiation takes place, and its products crystallise into different igneous silicate rocks. In the course of these processes heavy metals (Sn, W, Mo, Au, Ag, Pb, Zn, Cu, etc.) which are present in the magma in negligible quantities, and volatile constituents (H_2O, S, F, Cl, B, etc.) combine into readily soluble compounds, and on crystallisation of the magma, concentrate in the upper parts of the magmatic hearths. Sometimes the metals take part in the formation of residual silicate solutions the crystallisation of which gives rise to pegmatites containing minerals with F, B, Be, Li, Zr, occasionally with rare-earth elements. In other instances they emanate from the magmatic hearth in the form of gaseous products which have vigorous contact effects on the enclosing rocks with which they react chemically. Finally, as aqueous solutions (hydrotherms) they are carried through fissures into the roof of magmatic masses, where they often form rich deposits of metalliferous economic minerals.

Only a few heavy metals remain in the magma, and in the course of differentiation, concentrate in some of the rocks within the magmatic masses. When the magma erupts to the surface as lava, the volatile constituents escape into the atmosphere.

In accordance with the above sequence of development of the magmatic cycle, the following stages of endogenous processes of mineral formation are distinguished: (1) the magmatic stage proper; (2) the pegmatitic stage, and (3) the pneumatolytic-hydrothermal stage.

1. Magmatic processes had occurred during all geological eras and have given rise to tremendous masses of igneous rocks.

According to the conditions of their formation these rocks fall into two principal groups: (a) *effusive* (extrusive), i.e., those that reached the earth's surface as lava or rapidly solidified in direct proximity to the surface under low external pressure, and (b) *intrusive*, i.e., those that solidified at great depths under high pressure in the form of large mushroom-like, sheet-like or irregularly shaped blocks. Effusive rocks solidify rapidly, before crystallisation is completed and therefore contain varying quantities of volcanic glass. Often (in vesicular lavas) they also contain numerous spherical cavities formed through the evolution of gaseous products upon a sharp decrease in the external pressure. Intrusives, on the other hand, are holocrystalline (completely crystallised) rocks.

Magmatic differentiation, as was stated earlier, gives rise to rocks differing in chemical and mineral composition as well as in specific gravity. Depending on the content of silica and other constituents, igneous rocks are classed as:

(a) *Ultrabasic*, rich in MgO and FeO, but poorest in SiO (<45 per cent), e.g., dunites, pyroxenites in intrusive complexes, and pycrites in effusive complexes;

(b) *Basic*, richer in SiO_2 (45-55 per cent), rich in Al_2O_3 and GaO, but poorer in MgO and FeO, e.g., gabbro, and norites in intrusive complexes, and basalts and diabases in effusive complexes;

(c) *Intermediate* acidity rocks, SiO_2 content (55-65 per cent), poorer in CaO, but richer in alkalis: diorites, quartz diorites in intrusive complexes, and porphyrites, andesites, etc., in effusive complexes;

(d) *Acid*, rich in SiO_2 (> 65 per cent), still richer in alkalis, but poorer compared with the preceding ones in CaO, FeO, and MgO: granodiorites, granites, etc., in intrusive complexes, and liparites, quartz porphyrys, etc., in effusive complexes.

Somewhat apart stands the family of quartzless nepheline syenites (about 55 per cent of SiO_2), which are richer in alkalis and Al_2O_3 than granites, as well as of phonolites, leucitophyres, and other effusive complexes.

In a number of intrusive blocks where the differentiation of the magma was more complete, acid rocks are found at the top, while the ultrabasic and basic rocks, whose specific gravity is higher, in the deeper parts of the blocks.

Ore deposits of magmatic origin are found only in ultrabasic and basic igneous rocks. This includes deposits of Cr, Pt, and other metals of the platinum group, as well as Cu, Ni, Go, Fe, Ti, etc.

Alkali-rich intrusives (nepheline-syenites) may be associated with deposits of rare earths, such as columbium, tantalum, titanium, zirconium, and non-metallic mineral deposits, e.g., phosphorus (apatite), aluminiferous rock (nepheline), etc.

2. Pegmatite-forming processes take place in the upper marginal parts of deep-seated magmatic blocks (several kilometres from the surface of the earth) where external pressure is high enough to keep in solution the volatile magma constituents, which react with the earlier crystallised rock.

Pegmatites as geological bodies occur as veins or irregularly-shaped deposits, occasionally as stocks distinguished by unusually coarse-grained texture of the mineral aggregates. The pegmatite veins may be several metres thick, and on the strike they generally extend for tens, more rarely for a few hundred metres. For the most part pegmatite bodies are enclosed in the parent igneous rocks, but sometimes occur in the form of vein-like bodies in the invaded rocks.

It should be pointed out that pegmatite formations are found in intrusive rocks of most diverse composition, from ultrabasic to acid. Most commonly, however, pegmatites occur in acid and alkaline rocks. Those found in basic rocks are of no commercial importance.

In composition, pegmatites differ very little from parent rocks, being composed mainly of the same rock-forming minerals. Only minor (quantitatively) minerals which are not found in every type of pegmatites substantially differ in composition, since they contain valuable rare chemical elements often in association with minerals containing volatile constituents. Thus, besides the principal rock-forming minerals (feldspar, quartz, micas) granitic pegmatites contain minerals of fluorine and boron (topaz, tourmaline), beryllium (beryl), lithium (lithium micas), occasionally rare-earth minerals, columbium, tantalum, tin, tungsten, etc.

Many pegmatite formations show a zonal structure and

rather regular distribution of minerals. For instance, in the pegmatites of the Murzinka area in the Urals 51 the outer pegmatite zones at the contact with the enclosing granites are composed of light-coloured close-grained rock (aplite). Towards the centre, they are replaced by "graphic granite" zones (intergrown quartz and feldspar) followed by zones of coarse-crystalline feldspar and quartz. In the central zone of the pegmatite vein there are cavities ("bags") whose walls are lined with druses of large, well-formed crystals of rock crystal, topaz, and other gemstones.

The pegmatites which penetrate into intrusives very rich in alkali earths (MgO, CaO) sharply differ in mineral composition from pegmatites in the parent rocks. In that case, mineral paragenesis gives an indication of the vigorous reactions that have taken place between the solutions and the enclosing rocks. In such cases there are established mineral associations in which not only the elements of the magma (Si, Al, alkalis, etc.) take part but also the elements of wall rocks (MgO and CaO) which, at the contact with the pegmatites, also undergo sharp changes. Such pegmatites are referred to by A.Y. Fersman as "hybrid" pegmatites, as opposed to "purebred" pegmatites.

The origin of pegmatites has not yet been completely deciphered. According to A.Y. Fersman, they are the product of crystallisation of residual melts enriched in volatiles. Recently Soviet scientist A.N. Zavaritsky, proceeding from physico-chemical considerations, suggested the possibility of formation of coarse-crystalline masses through crystallisation of parent rocks under the influence of gases accumulating in the residue of the crystallisation of the magma. However, in all cases the pegmatites are formed towards the end of the magmatic process proper and, as such, occupy an intermediate position between plutonic magmatic rocks and hydrothermal ore deposits.

3. Pneumatolytic-hydrothermal processes are essentially post-magmatic, i.e., they occur when the main process of magma crystallisation in a deep-seated block is over in the main.

Pneumatolysis (*pneuma* is the Greek for gas) may occur when melts saturated with volatile constituents, crystallise under the conditions of relief of external pressure. Therefore at a certain pressure the residual melt boils up, as it were, passing into gas which interacts with the formed earlier solid minerals. In other words, there occurs the process of sublimation. Processes of this kind occur when magmas solidify at great or medium depths.

In this case the volatiles chemically react with the enclosing rocks, which gives rise to *contact metamorphism.* This is accompanied by chemical reactions in the wall rocks (the roof) *impregnated with solutions.* The degree of metamorphism and the composition of the resulting products depend not so much on temperature, as on the chemical activity of the solutions and the composition of the rocks with which they react. The most profound changes are observed in *limestones* and other calcareous rocks contacting the magmatic masses. Chemical reactions, involving the mechanism of replacement or metasomatosis, give rise to so-called *skarns* which are mainly composed of silicates of Ca, Fe, Al, etc.

Their chemical composition indicates that skarns have derived both from the wall rocks (limestones, dolomites, etc.) and the magma constituents. It should be noted that the intrusives which have already solidified by the time the process under consideration takes place, also alter along the contact. In this process the minerals of magmatic rocks are replaced by new formations whose composition shows that an influx of elements (Ca, Mg) from the carbonate strata had taken place. Skarns are often associated with deposits of iron, occasionally of tungsten, molybdenum, and other metals.

When the magmas erupt to the earth's surface, enormous quantities of volatiles are discharged into the atmosphere. However, in the fissures of cold lavas, on the walls of volcano craters, and in other enclosing rocks one often observes sublimation products such as native sulphur, ammonium chloride, borates, etc. Some metasomatic reactions are also observed, but they are less pronounced than in the previous case.

Hydrothermal processes occurring at great depth develop in the roof, at some distance from the direct contact with igneous rocks. The residual vaporous solutions which move through a network of fissures developing in the roof of the magmatic hearths in the course of magmatic intrusion gradually condense into hot aqueous solutions or hydrothermas.

Conditions for hydrothermal processes are most favourable at low and medium depths (3 to 5 km) from the surface of the earth. The majority of hydrothermal formations are spatially and genetically associated with acid intrusive rocks $BaSO_4$, AS (granites, grandiorites, etc.).

The range of circulation of the solution extends sometimes from the upper parts of magmatic hearth right to the earth's surface. In areas of recent volcanic activity one still finds hot mineralised springs depositing siliceous material with substantial amounts of sulphides of Hg, Sb, As, Pb, Cu, etc.

Farther away from the magmatic hearth, the hydrothermal solutions enter an environment which is gradually enriched in oxygen. At the same time there is a decrease in external pressure and temperature, the latter diminishing from approximately 400° to a few tens of degrees. These factors, naturally, influence the course of chemical reactions and the mineral composition of the hydrothermal solutions.

By the token of the prevailing mineral associations, these formations are tentatively classified as high, medium, and low-temperature ones. This does not mean, of course, that high-temperature formations cannot be found among mineral associations crystallising at lower temperatures. Even in pegmatites and contact-metamorphic formations there are invariably found lower-temperature minerals of hydrothermal origin. These, however, reflect the final stages of the process of mineral deposition which began at high temperatures.

The formation of hydrothermal solutions continues for a very long time, evidently as long as the magmatic hearth exists. Soviet scientist S.S. Smirnov deduced from an analysis of the correlation of different deposits forming a single ore complex that the movement of the mineralising solutions must have been intermittent due to repeated resumption of fracturing processes. This is borne out by the frequent indications of superimposition of later mineralisation stages on earlier ones.

The shape of the mineral bodies depends on the configuration of the cavities filled by them and, to some extent, on the composition of the rocks through which the solutions had circulated. When fissures had been filled, this results in interrupted veins the roots of which sometimes go down to the upper parts of the magmatic masses. Deposition of minerals in tiny pores and openings produces *impregnations*. The passing of the solutions through chemically reactive rocks (e.g., limestones) gives rise to *metasomatic deposits* which often are irregularly shaped. Should a solution suddenly encounter a large opening, the sharp pressure relief causes intense evaporation of the solvent (water) and hence, at least initially, considerable supersaturation of the solution and the precipitation of colloids. Indeed, metacolloidal formations on the walls of veins are quite common, particularly when the processes were associated with shallow-seated intrusives. Also very common are cavities lined with druses of different crystals.

The mineral composition of hydrothermal deposits is extremely varied. The veins in the vast majority of cases are composed of quartz containing diverse minerals, mostly metal sulphides. It should be noted that it is hydrothermal deposits that produce most of the ores of rare (W, Mo, Sn, Bi, Sb, As, Mg, sometimes Ni, and Co), non-ferrous (Cu, Pb, Zn), precious (Au and Ab), and radioactive (U, Ra, Th) metals.

Processes occurring on the earth's surface which derive their energy from the heat of the sun are more accessible to observation than endogenous processes. On dry land, atmospheric agents (oxygen, carbon dioxide, water) and bacterial activity give rise to a powerful chemical process known by the general name of *weathering*. It consists in the physical and chemical disintegration of everything that had been formed by the endogenous processes and at the same time gives rise to products which are stable in the new conditions on the surface of the earth.

In part, these products are transported by running surface waters either in solution or in suspension and deposited wherever the current becomes more sluggish, in river valleys, or in lacustrine and marine basins. The latter become the site of specific processes of mineral formation resulting in the deposition of sediments in the form of beds on the bottom of the basins. This is known as the *sedimentary* process.

1. Weathering in the first place takes the form of mechanical destruction of rocks and ores caused by temperature fluctuations, as well as by the action of water freezing in cracks and pores, and by other factors. All this results in the disintegration of rock and ore minerals which possess different expansion factors. Much more important, however, is the chemical weathering of minerals by rain and surface waters which contain dissolved oxygen, carbon dioxide and other gases, and therefore possess rather strong oxidising and dissolving properties. Percolating to the ground water table, these waters gradually lose their oxygen in the course of reactions of oxidation, hydration, and carbonation.

The leaching-out of the soluble compounds thus formed leads to the formation of pores, caverns, and sometimes even great chambers (karsts). The walls of such chambers are often covered with colloidal sinter formations or crystal druses of exogenous minerals, or else may be coated with earthy ochreous matter. Where intense leaching of more or less readily soluble rocks takes place, the soil over them sags forming sinks and sometimes swallow holes, or even great caves (in gypsum and limestone series).

Vegetation and various organic compounds which pass into solution intensify considerably the chemical weathering of ores and rocks.

Chemically inert minerals (quartz, gold, platinum, etc.) and also recent formations of low solubility accumulate in the residual products at the surface in the form of clayey masses of different light and dark shades, but most often stained brown by iron hydroxides.

Accumulating at or near the surface, the insoluble products of chemical weathering form the so-called *residual deposits* composed predominantly of hydroxides and hydrosilicates. Examples are the numerous deposits of clays, kaolins, and bauxites, as well as of iron, nickel, and other ores, formed in the course of intense disintegration of rocks of corresponding composition. These deposits have a vast areal extent.

The chemical weathering of economic mineral deposits gives rise to residual formations called *caps* (iron, manganese, gypsum, and others). In the case of iron, they are known as iron hats or gossans. Because certain constituents are leached out, the concentration of residual economic minerals in these caps is generally much higher than in the undecomposed primary ores, i.e., those lying below the ground-water table. It should be noted that the leached-out metals, especially copper, and also silver and zinc which percolate in aqueous solutions down to the lower part of the zone of oxidation, i.e., to the ground-water table, react with primary ores or with the chemically active wall rocks (limestones). Thus, a zone of *secondary* sulphide enrichment with a much higher concentration of copper in the ores develops in copper-sulphide deposits.

Climatic factors (mean annual temperatures and precipitation) greatly affect chemical weathering. Low humidity and high mean annual temperatures accelerate the oxidation and concentration of chemical compounds. Topography is also an important factor. In mountainous regions intense erosion carries away the products of chemical weathering before they can form large accumulations. A different picture is observed in flat country.

The deposits formed by weathering commonly have somewhat irregular pocket- or sheet-like shapes which are more or less parallel to the daylight surface. Along large fissures, zones of crushing, and the contacts of physically and chemically different rocks, i.e., where the surface weathering agents penetrate to greater depth, steep ore bodies of surface origin, pinching out with depth, may develop.

2. Sedimentary processes occur in aqueous environment: rivers, lakes, and seas. In sea basins these processes through geologic time gave rise to the formation of series of sedimentary rocks of enormous thickness. Sedimentary rocks maybe divided into mechanical and chemical.

Mechanical sediments are formed as a result of the erosion of the weathering products and the re-deposition by running water of the chemically inert minerals and rock fragments such as pebbles, gravels, sands and sand clays in river valleys and aquatic basins. If the eroded products of weathering contain chemically inert valuable minerals, they are re-washed and redistributed in river valleys in accordance with their specific gravity, forming *placer deposits* which often are of commercial importance. Such are the gold, platinum, diamond, and some other placers.

No new minerals are formed in the process of accumulation of mechanical sediments. Only in the case of ancient placers does one observe later chemical alterations in the clastic material.

Chemical sediments are mostly limited to lacustrine and marine basins. They are formed by crystallisation of saturated salt solutions, by deposition of coagulating colloids in the form of gels, and finally, by accumulation of the products of activity of the organic kingdom and of the organic remains themselves.

(1) The formation of *crystalline* sediments occurs in many

drying lakes in regions with an arid and hot climate where surface evaporation prevails over influx of fresh water.

Crystallisation begins at a certain super saturation of the aqueous solutions. The order of the segregation of minerals with progressive evaporation of water depends on the two main factors of the equilibrium of the system: the composition of the solutions or, to be more precise, the ratio of concentration of the components forming the system, and the temperature of the solutions at which crystallisation takes place. The equilibrium conditions of sulphates and chlorides of Ca, Mg, K, and Na in sea water have been thoroughly investigated for different concentrations and temperatures by N.S. Kurnakov and many others.

(2) The formation of *colloidal* sediments in lacustrine and marine basins is much more complex, and some of its aspects are still obscure. Thus, certain compounds which result from weathering are transported by running water, not only in true solutions but also in colloidal solutions or sols which are stable in fresh water. When these solutions are transported by surface waters into marine basins, they are coagulated by the abundant electrolytes present in sea water in the form of ions of the dissolved salts. This is the case with colloidal solutions of iron, manganese, silica, and other oxides.

The gels, thus formed, together with the clayey and fine detrital particles brought by fluvial waters as well as the remains of marine life are deposited at the bottom of the near-shore zone, either as intercalations or thicker and more regular beds. With time, these sediments undergo alteration (diagenesis) turning into compact masses.

Sedimentary manganese deposits offer a good example of the dependences governing the alteration of paragenetic associations of minerals due to the physico-chemical conditions in which the sediments are formed at the bottom of basins.

Tetravalent manganese compounds which are the richest in oxygen are prevalent in shallow near-shore zones. Farther away from the coast they are replaced by divalent manganese carbonates accompanied by rare iron sulphides.

Sedimentation in the shallow zones apparently took place in the presence of the oxygen, which down to a certain depth is dissolved in sea water, whereas in the deeper zones, in conditions of oxygen deficiency, the carbonates and the accompanying sulphides apparently formed through the action of carbon dioxide and some hydrogen sulphide evolved in the course of decomposition of organic remains.

Hence there emerge the so-called *fades* of ores of different composition (oxides and carbonates). A similar relationship between sediments of different composition apparently exists in deposits of iron ores, for which facies of oxides, silicates, and carbonates have long been known. As to the processes occurring in deeper parts of marine basins and oceans, our knowledge is still very scarce.

(3) Organogenic or *biogenic* sediments which are formed as a result of a very complex biological process include limestones consisting of skeletons of marine animals, diatomites consisting of siliceous skeletons of diatoms, and caustobioliths (*caustos* means combustible in Greek) formed chiefly from vegetal and partly from animal organisms (fossil coals, shales, oils, combustible gases, solid bitumens, etc.).

Organogenic sediments may develop through the accumulation of the skeletons of dead animals (coquina) or from the tissues of higher or lower plants (peat, sapropel). They also may result from the life activity of the organisms themselves, e.g., the decomposition by anaerobic bacteria of organic remains or sulphates, giving rise to sulphur deposits. Finally, laboratory experiments with ferrobacteria have shown that the latter may give rise to nodular formations.

On subsequent transformation, some of these sediments

turn into inorganic products (limestones, phosphorites), while others remain organic (coals, etc.).

Regional Metamorphism and Related Mineral Formation Processes

The most profound changes both in endogenous and exogenous formations occur in the process of so-called regional metamorphism when as a result of tectonic displacement whole areas of the upper parts of the earth's crust sink to great depths, finding themselves in the conditions of high temperatures and pressures or the conditions of powerful orogenic processes.

In this new environment, the mineral and chemical composition, physical properties and external appearance of the rocks and ores undergo substantial alteration. Exogenous water-rich compounds become partially or completely dehydrated, (e.g., opal changes to quartz, limonite to haematite or magnetite, etc.). This is accompanied by recrystallisation (organogenic limestone, for instance, turns into marble with the obliteration of its former structural characteristics).

In many rocks, including igneous ones, a complete rearrangement of the constituents occurs with the formation of new minerals. Some minerals, for instance, gypsum, native sulphur and rock salt never occur in metamorphic series. Under the action of high temperature and pressure, chemical reactions tend to produce minerals of smaller volume and higher specific gravity. Mineral paragenesis depends not only on the composition c the rocks undergoing metamorphism but also to a large extent on the depth at which this occurs, i.e., on the thermodynamic conditions.

The rocks themselves under high dynamic stress change to schists which readily split into plates (clay shales, slates, mica schists, gneisses, etc.). If metamorphism affects finely laminated sedimentary rocks so that the stress applied roughly coincides with the direction of lamination, the laminations will be crumpled into numerous fine folds.

There is no doubt that such constituents as H_2O, CO_2, and other mineralisers which promote not only recrystallisation, but metasomatosis and even re-deposition of the minerals play a large part in the rearrangement of mineral matter. In this case H_2O and CO_2 originate either from magmatic rocks or rocks undergoing metamorphism. Some, especially sedimentary rocks, in the course of their recrystallisation into aggregates of anhydrous minerals should release large quantities of water and some carbon dioxide.

Under high temperatures and pressures, the metamorphic water should acquire all the properties characteristic of hydrothermas which are genetically associated with magmatic intrusion, i.e., higher dissolving capacity and ability to carry and re-depose minerals either along fissures or through metasomatosis. This, however, does not exclude the possibility of the impregnation of metamorphosed rocks with magmatic water vapours, especially in regions of great intrusions of acid igneous rocks.

Among economic mineral deposits occurring in metamorphosed rocks the following genetic types are identified: (a) *metamorphosed* deposits which existed before metamorphism began (e.g., sedimentary deposits of iron and manganese), and (b) *metamorphic* deposits, i.e., those produced in the metamorphic process.

The latter type is exemplified in the formation of graphite in metamorphic rocks from organic remains. Sometimes cryptocrystalline graphite is found, bearing the imprints of plants deriving from coal seams (in metamorphic series of the eastern slope of the Urals mountains). The graphite, in this case, is no longer a combustible substance due to a radical change in its original properties and the loss of its volatile constituents.

This genetic type also includes the so-called *Alpine* clefts (the name referring to the place of its discovery) which are very interesting mineralogically. The veins have long since attracted the interest of mineralogists because of the beautiful druses of crystals of different minerals found in them. They are associated with fissures in metamorphic rocks, usually across the bedding plane. The important feature of their composition is that the minerals crystallising in them are exactly the same as those which form in the course of metamorphism in the *enclosing* rocks.

They are also found in approximately the same quantitative ratio. Only the more soluble rare minerals containing Ti, P, Cl and B are present in fissures in greater quantities than in the wall rocks. Another important feature is that these fissures generally do not contain chemical elements or minerals that might point to any genetical relationship with the hydrothermal activity of acid magma intrusions (e.g., minerals of gold, silver, lead, zinc, antimony, tin and tungsten), although the possibility is not excluded that magmatic water too takes part in these processes.

Thin cracks in metamorphic rocks are completely filled with mineral matter, examples being the veins of white calcite in grey limestones, of milk-white quartz in red jaspers, etc. It should be noted that the crystal grains in these veinlets are always larger than those in the host rock.

6

The Petrogenesis

The term petrogenesis refers to the means by which a rock or mineral deposit is formed. Three different modes of formation are common; these consist of igneous activity, metamorphism, and sedimentary processes. Mineral deposits may thus occur in igneous, metamorphic or sedimentary environments and also in vein environments.

Igneous rocks may also be classified according to their texture, a characteristic defined by the size and shape of the grains of which they are composed. The size of the crystals in an igneous rock depend largely on the rate of cooling of the magma which formed the rock. Plutonic rocks tend to have large crystals because magma trapped at depths within the earth cools slowly. Magma which is ejected by a volcano, on the other hand, tends to cool quickly. Volcanic rocks therefore tend to contain only very small crystals.

One final means of classifying igneous rocks is by composition. Rocks are divided into two classes according to whether they contain a greater proportion of iron and magnesium or of potassium, sodium, and calcium. Igneous rocks which contain a high percentage of iron and magnesium tend to possess a dark colour and are said to be of basaltic or mafic composition. In contrast, igneous rocks which contain a greater percentage of potassium, sodium, or calcium tend to have a lighter colour and are said to be of granitic or felsic composition. Rocks with an intermediary composition are termed andesitic.

Metamorphism describes the set of solid state processes which transform one type of rock into another. Any type of rock, whether igneous, metamorphic, or sedimentary, may be metamorphised. Metamorphic changes occur mainly in mineral structure and texture; changes in chemical composition may also take place. Metamorphism typically occurs at the high temperatures and pressures which are present deep within the earth's crust. The process is defined to occur in the solid state and melting of the rock may therefore not occur. Any process during which complete melting of the rock does occur is considered to be of igneous rather than metamorphic nature.

Metamorphic processes may be divided into two categories according to the area of their extent. Regional metamorphism consists of geologic processes which act over wide areas and cause metamorphic changes in vast expanses of rock. In contrast, local metamorphism affects only relatively small areas. Local metamorphism occurs when an igneous body or pluton intrudes into and heats a surrounding body of rock. For this reason local metamorphism is also called contact metamorphism.

Different types of metamorphism may alternately be categorised according to the agents which are active during each process. Stress, pressure, and shock are causative agents of dynamic metamorphism, while thermal metamorphism is due merely to the influx of heat from a nearby igneous intrusion. In contrast, hot, volatile fluids play a significant role in hydrothermal metamorphism. Hydrothermal metamorphism takes place when volatile solutions percolate into and react with a host rock. Both the heat of the intrusive igneous body and the reactive fluids serve to catalyse metamorphic reactions in the host rock. The intrusive magma and the associated volatile fluids may introduce previously absent elements into the reaction process; the chemical composition of the host rock may therefore be extensively altered during hydrothermal metamorphism.

A third possible method of classifying metamorphic processes is whether or not previously absent elements are introduced into the system. Isochemical metamorphism consists of the recrystallisation of previously present elements into new species of minerals. Such minerals may differ structurally from the original minerals, but will contain only those elements which were originally present. In contrast, allochemical metamorphism introduces new elements into the system during the reaction process and the resultant mineral species contain elements not previously present. Dynamic and thermal metamorphism are thus types of isochemical metamorphism, whereas hydrothermal metamorphism is an allochemical process.

Veins are mineral deposits which form when a fracture or fissure within a larger body of rock is filled with new crystalline material. Veins are believed to form when aqueous solutions migrate through fissures in rock and deposit minerals onto the fissure walls.

Most veins form as new mineral species are precipitated onto rock walls but leave the wall rock unaltered. In such cases minerals fill the original crack or fissure but do not extend into the host rock, and the boundary between host rock and newly deposited vein minerals remains clearly delineated. Because the mineral species which compose the veins were precipitated by hot water, vein deposits of this nature are a type of hydrothermal deposit and the mineral material of such veins is chemical sedimentary rock. However, occasionally the rock wall which contains the vein undergoes alteration. Portions of the host rock may dissolve or react chemically with the circulating volatile fluids. In this case the boundary between vein deposit and original rock wall will be unclear. Such veins are termed hydrothermal replacement deposits. Hydrothermal replacement deposits are a form of hydrothermal metamorphism.

Sediment includes all uncemented, particulate matter which accumulates at the earth's surface. Sediments are derived from the erosion of pre-existing rocks, by precipitation from aqueous solution, or from the skeletal remains of organisms, and are transported by flowing water, wind, and moving glaciers. Sedimentary rock is formed when loose sediments are collected together, compacted, and cemented into rock with the aid of heat, pressure, and chemical agents.

The term diagenesis is used to portray the processes which result in the conversion of loose sediment into sedimentary rock. These changes include recrystallisation; lithification, or compaction and cementation; and precipitation. Compaction is a physical change which occurs as layers of sediment accumulate in one area and the lowest layers are gradually compressed by the weight of the layers above. Cementation, in contrast, is a chemical change during which aqueous solutions seep through the sediment, filling spaces and cementing individual particles together. Sedimentary diagenesis takes place at relatively shallow depths and low temperatures.

Sedimentary rocks may be classified according to their texture. Such rocks are categorised according to whether they are clastic or non-clastic. Rocks of clastic texture exhibit a mass of discrete particles and grains which are cemented together but retain their individual shape, whereas rocks which are non-clastic do not contain visible, macroscopic grains or particles.

Sediments are derived by the two different processes of mechanical and chemical weathering. These two different means of derivation provide another method for the classification of sedimentary rocks. Detrital sedimentary rock contains particles and sediment which were derived by mechanical means. The materials which compose such rocks were ground down by mechanical processes such as the passage of wind, the movement of glaciers, and splitting due to freezing and thawing. These materials were then transported to the site of deposition by mechanical means such as wind, flowing water, or travelling glaciers. Few chemical changes occur in the mineral material of which detrital sedimentary rocks are composed. Chemical sedimentary rock, on the other hand, contains only substances which were derived by chemical weathering processes. The constituent materials of chemical sedimentary rock were dissolved in water, transported to the deposition site in the form of aqueous solutions, and finally precipitated.

Igneous Processes

Magma is molten rock which originates deep within the earth. Magma is formed as rock melts at the intense temperatures and pressures present at great depths within the earth. Igneous rocks form when magma rises towards the earth's crust, cools, and solidifies. The name igneous is derived from ignius, a Latin term meaning 'fire'. Most of the rocks which compose the earth, moon, and other planets are igneous in origin.

Melting of Rock

It is hypothesised that the earth's crust and mantle are composed mainly of solid rather than molten rock. It is also estimated that rock within the lower crust and upper mantle maintains a temperature of 1200° to 1400° Celcius. This temperature is very close to the melting point of rock. With the aid of added heat, volatile agents, or a change in pressure, melting may occur.

One potential source of heat is friction in subduction zones as one crustal slab is subsumed by another. Another is the introduction of heat by mantle rock rising from yet greater depths. Either of these may serve to introduce the heat necessary in order to initiate partial melting of the mantle rock. Because an increase in pressure causes an increase in the melting temperature of rock, rock may also undergo a solid to liquid phase transition if the pressure of its environment drops. This process is termed as 'decompression melting'. Rock which contains a significant portion of volatile fluids such as water vapour (H_2O) and gaseous carbon and sulphur dioxide (CO_2 and SO_2) has a lower melting point than rock which lacks such volatiles. If volatiles are introduced into previously solid rock, melting may be induced. Frictional heat, externally introduced heat, volatile fluids, and the reduction of pressure may thus all serve as agents which initiate the partial melting of mantle rock and the formation of magma.

Ascent and Evolution of Magma

When a magma is less dense than the surrounding mantle rock, the rock exerts an upward force on it. Due to this buoyant force the magma then rises from its place of origin deep within the crust. Both the viscosity, qualitatively defined as the internal resistance to fluid flow, and the density of a magma determine how fast it rises. A large silica content tends to increase the viscosity of a magma. In contrast, the presence of volatile agents such as H_2O, CO_2 and SO_2 tends to decrease viscosity. An increase in pressure will decrease the viscosity of a magma but will increase its density.

An abundance of silica (SiO_2) tends to increase the occurrence of polymerisation, or the formation of silica chains. A more polymerised magma possesses a greater viscosity and therefore flows more slowly towards the surface of the earth. On the other hand, the presence of volatile fluids such as H_2O, CO_2 and SO_2 may decrease the viscosity of a magma and thereby speed its flow. Because magma is much more compressible than crystalline substances, its density rises at high pressures whereas its viscosity decreases. A change in pressure may therefore either speed or slow the magma's ascent towards the surface.

Magma is not typically a simple liquid. Instead, it is usually a mixture of molten rock, solid crystals, and dissolved gases. Most magmas carry volatile agents such as H_2O, CO_2 and SO_2. As a magma rises through the earth's crust, its chemical composition may undergo a series of changes. Crystallisation may remove some elements from solution and increase the concentration of those which remain. Melting of the enclosing rock may introduce into the solution elements which were previously absent. Through processes such as these a magma will change in composition as it rises through the crust.

Minerals may crystallise within the magma and then drop out of solution under the effect of gravity; alternately, crystals which are less dense than the magma may float to the upper portions of the enclosing chamber. Crystals may form along the enclosing rock walls and be left behind as the magma ascends through the crust. Components of the magma may exsolute, crystallise, and be carried along as the magma rises. All such crystallisation changes the constituency of the magma by removing certain elements from solution and increasing the concentration of those which remain.

The magma may also break off portions of the wall rocks through which it passes and transport them towards the earth's surface. Minerals from the rock wall may be melted and assimilated into the magma. Both of these possibilities may introduce to the solution elements which had previously been absent. Minerals which formed earlier may continue reacting with the still-fluid portion of the magma to form new substances. Through processes such as crystallisation and melting of surrounding rock a magma may change dramatically in composition as it rises through the crust. A wide variety of igneous rocks may thus originate from a single parent magma.

Plutonic Versus Volcanic Rocks

Igneous rocks are classified according to whether they formed from magma which cooled slowly at great depths within the earth's crust or from magma which erupted at the earth's surface. Intrusive igneous rocks, or rocks which solidify while buried deep within the earth, are called plutonic rocks. This term was derived from the name of Pluto, Greek god of the underworld. In contrast to intrusive or plutonic rocks, extrusive rocks form from lava, or magma which erupts at the earth's surface before cooling. Extrusive igneous rocks are termed volcanic rocks, after Vulcan, the Roman god of fire. Intrusive rocks crystallise slowly at high temperatures and pressures whereas extrusive rocks crystallise at atmospheric pressure and cool quickly.

The size of the crystals which compose an igneous rock depend largely on the rate of cooling of the magma which formed the rock. In a magma which cools quickly, large crystals do not have time to form, and the resultant rock will be glassy or contain only tiny crystals. In contrast, a magma which pools at great depths and losses its heat slowly may develop crystals which are quite large.

Once a lava erupts at the earth's surface it tends to loose heat very rapidly to its surroundings. Because they usually cool quickly, volcanic rocks are typically of glassy texture or contain only very tiny crystals. Plutonic rocks, in contrast, solidify from magmas which pool at great depths within the earth's crust. At such depths the surrounding rock wall is already at a relatively high temperature. Plutonic rocks therefore tend to loose heat only very slowly to the rock wall which surrounds them. Large plutons - which may possess a diameter on the order of 100 kilometres - can take hundreds of thousands of years to solidify. Because such rocks cool very slowly, they tend to form large crystals. Plutonic rocks are observed at the earth's surface due to erosion or tectonic upheaval long after crystallisation.

Texture

The term texture describes the size and shape of the grains which compose a rock as well as the manner in which they are arranged. Igneous rocks may be classified according to their texture. Volcanic rocks cool and crystallise quickly from magma which has been extruded upon the earth's surface. During such rapid cooling large or complex crystalline structures do not have time to form. Volcanic rocks therefore tend not to contain visible crystals or else to contain only tiny ones. The texture of such rocks is termed glassy or aphanitic. Plutonic rocks, in contrast, cool and crystallise slowly at great depths and may thus contain large crystals. The texture of these rocks is described as phaneritic. Porphyritic texture describes a rock which contains a few large crystals embedded in an aphanitic matrix.

Glassy Texture

Some igneous rocks formed as lava erupted at the surface of the earth and cooled before a crystalline structure formed. Such a rock possesses a disordered internal structure and is called a glass. Obsidian is a familiar example of a volcanic rock which demonstrates glassy texture.

Aphanitic Texture

Rocks which cooled quickly at or near the surface of the earth contain only very small crystals. The texture of such fine-grained rocks is termed aphanitic, which is derived from the Greek aphanes, or 'invisible, unseen'.

Phaneritic Texture

In contrast to rocks of aphanitic texture, igneous rocks of phaneritic or 'visible' texture contain macroscopic crystals. Phaneritic rocks are of plutonic origins; in order to enable the growth of macroscopic crystals the process of cooling and crystallisation of magma had to take place at a relatively slow rate deep within the earth's crust.

Porphyritic Texture

Because crystals of differing chemical compositions solidify at different rates, it is possible for certain crystals to reach significant size before others have formed at all. Occasionally a magma which contains a few large crystals but is otherwise still liquid shifts suddenly towards cooler surface temperatures. In such cases the crystals may be transported along with the liquid portion of the magma.

The remaining magma would then crystallise quickly at the new, cooler surface temperatures. The resulting rock may contain a few large crystals embedded in an aphanitic matrix of tiny, quickly-cooled crystals. This type of texture is termed porphyritic.

Pegmatites

A pegmatite is a body of plutonic rock which contains unusually large crystals. Crystals measuring thirty centimetres across are common. Quartz crystals weighing thousands of kilograms and mica crystals of three metres or more in diameter have been found in association with pegmatites.

Pegmatites often extend outwards as veins or dikes from larger masses of plutonic rock. The formation of a pegmatite is probably directly related to the formation of this larger plutonic body.

As the magma within the main chamber cools, minerals crystallise and exsolute, increasing the concentration of volatiles such as water vapour (H_2O), boron (B), fluorine (F), chlorine (Cl), and phosphorous (P) in the remaining magma. Higher concentrations of such volatiles increase the fluidity and lower the viscosity of the magma which remains, which may then escape the main chamber and penetrate into the surrounding rock as a vein or dike. In this more fluid magma, ions possess an unusually high mobility, and atypically large crystals may develop as the magma cools.

Pegmatites are occasionally rich in molybdenum, lithium, cesium, uranium, and titanium-bearing minerals. Gemstones such as garnet, tourmaline, beryl, and topaz have been found in pegmatite environments.

Composition of Igneous Rocks

The chemical composition of a magma at the time when it cools determines the identity of the minerals which crystallise from the magma and therefore the identity of the resultant igneous rock. The most prevalent component of magma by weight is typically silica (SiO_2). However, magma also contains in varying quantities ions of all the other elements (aluminium, Al, iron, Fe, calcium, Ca, sodium, Na, potassium, K, and magnesium, Mg) which compose the bulk of the earth's crust.

As a magma cools its constituent elements bond to form two different types of silicate minerals. These two types of silicates are divided according to which metallic elements they contain. The ferromagnesian silicates are rich in iron and magnesium, although they may also contain sodium or calcium. These include olivine (Mg_2SiO_4, Fe_2SiO_4), the minerals of the pyroxene group - enstatite, $Mg_2(Si_2O_6)$, hypersthene, $Fe_2(Si_2O_6)$, diopside, $CaMg(Si_2O_6)$, and hedenbergite, $CaFe(Si_2O_6)$-and the minerals of the amphibole group- anthophyllite, $(Mg,Fe)_7(Si_8O_{22})(OH)_2$, tremolite, $Ca_2Mg_5(Si_8O_{22})(OH)_2$, actinolite ($Ca_2Fe_5(Si_8O_{22})(OH)_2$), and glaucophane, $Na_2Mg_3Al_2(Si_8O_{22})$, as well as biotite mica ($K(Mg,Fe)_3(AlSi_3O_{10})(OH)_2$). In contrast, the non-ferromagnesian silicates contain potassium, sodium and calcium rather than iron or magnesium. These include quartz (SiO_2), the minerals of the feldspar group (orthoclase, $K(AlSi_3O_8)$, albite $Na(AlSi_3O_8)$, and anorthite ($Ca(Al_2Si_2O_8)$)) and muscovite mica ($KAl_2(AlSi_3O_{10})(OH)_2$).

Igneous rocks which contain a high percentage of the ferromagnesian silicates tend to possess a dark colour. In contrast, those igneous rocks which contain a greater percentage of non-ferromagnesian silicates tend to have a lighter colour.

Igneous rocks can be divided into two classes according to their proportional content of ferromagnesian and non-ferromagnesian silicates. Igneous rocks composed mainly of dark, ferromagnesian silicates are said to be of basaltic composition. They are also called mafic rocks, the word mafic being derived from the first syllables of magnesium and ferrum, or iron. Because of their iron content such rocks tend to be both denser and darker in colour than those rocks composed mainly of non-ferromagnesic silicates. Igneous rocks which are composed mainly of the light-coloured, non-ferromagnesian silicates such as quartz and feldspar are said to be of granitic composition. Granitic rocks are also called felsic rocks, the word felsic being derived from the initial syllables of feldspar and silica (or quartz). Such rocks tend to contain a relatively greater percentage of silica (SiO_2); typically this is about 70 per cent by mass. Rocks with an intermediary composition are termed andesitic after the volcanic rock andesite.

Carbonatite Deposits

Carbonatite is a unique type of intrusive igneous deposit which is rich in calcium carbonate (calcite, $CaCO_3$) and minerals of the carbonate class. Calcium carbonate is typically burned out of igneous materials by intense heat; the preponderance of calcite found in carbonatite therefore makes it a very unusual type of igneous formation. Carbonatite deposits may form as plugs, dikes, sills, or veins when carbonatite magma intrudes into host rock formations. Carbonatites are only observed to occur in continental plates and do not occur in oceanic plates or at plate boundaries.

Carbonatite magma differs greatly from the more prevalent silicate magma, and the two different types of magma are immiscible. Carbonatite rarely contains more than 10 per cent silica by mass and may contain much less. Such magmas possess a very low viscosity because the lack of silica prevents extensive silicate polymerisation. It has been shown experimentally that at a temperature of around 600° Celsius a magma possessing a high carbon dioxide (CO_2) content will divide into separate, immiscible silicate and carbonatite magmas.

One possible method of carbonatite formation is that a parent magma originating in the mantle underneath a continental crust rises until it reaches the boundary between crust and mantle. The magma may then be of a higher density than the crustal plate and may be detained. Ferromagnesic silicates may crystallise out at these high temperatures although non-ferromagnesic silicates remain liquid. Portions of the crustal plate, which contains plentiful carbonates, may also melt and be incorporated into the magma.

When the ferromagnesic silicates are removed from the magma its density decreases due to the newly lessened relative concentration of iron and magnesium. The magma may then rise into the crust until it reaches zones of lower temperature (around 600° Celsius) where it may separate into silicate and carbonatite magmas.

Carbonatites contain atypically high concentrations of rare earth elements such as titanium (Ti), vanadium (V), manganese (Mn), iron (Fe), copper (Cu), zinc (Zn), yttrium (Y), zirconium (Zr), niobium (Nb), molybdenum (Mo), barium (Ba), lanthanum (La), cesium (Ce), samarium (Sm), europium (Eu), lead (Pb), thorium (Th), and uranium (U). The elements sulphur (S), phosphorous (P), and fluorine (F) are also frequently present. High concentrations of magnetic materials may result in observations of unusual magnetic phenomena; atypically high concentrations of radioactive elements may produce unusual levels of radioactivity in the vicinity of a carbonatite.

Calcite ($CaCO_3$), dolomite ($CaMg(CO_3)_2$), and ankerite ($CaFe(CO_3)_2$) are the most common minerals found in carbonatites. The carbonates strontianite ($SrCO_3$), and magnesite ($MgCO_3$); the sulphides pyrite (FeS_2), molybdenite (MoS_2), galena (PbS), chalcopyrite ($CuFeS_2$), and sphalerite (ZnS); the oxides ilmenite ($FeTiO_3$), haematite (Fe_2O_3), rutile (TiO_2), and zircon ($ZrSiO_4$); the sulphate barite ($BaSO_4$); the phosphates monazite ($(Ce,La)PO_4$), and fluorapatite ($Ca_5F(PO_4)_3$); and the halide fluorite (CaF_2) are other species which are often present.

METAMORPHISM

Metamorphism describes the set of solid state processes which transform one type of rock into another. Any type of rock, whether igneous, metamorphic, or sedimentary, may be transformed into a metamorphic rock.

Metamorphism occurs when the protolith, or parent rock, is either subjected to temperatures and pressures greater than those at which it formed or else is chemically altered by incident fluids. The metamorphic process occurs as the minerals of the original rock become unstable at the new set of temperatures and pressures or upon the inclusion of previously absent chemical substances. During metamorphism, mineral and rock deposits change and adapt until equilibrium with the new environmental conditions is reached.

Agents of metamorphism are heat, volatile fluids, and pressure or stress. Heat capable of catalysing a metamorphic reaction may be supplied by nearby igneous intrusions, the latent heat of crystallisation of a nearby solidifying magma, or by the geothermal gradient within the earth's crust. Volatile fluids often act as a catalyst, reactant, or product during metamorphism. Such fluids are typically a mixture of H_2O, CO_2 and SO_2 and may be derived from meteoric groundwater, seawater, or a nearby igneous intrusion. Pressure and stress applied to rock may also induce metamorphism.

Metamorphic changes occur mostly in mineral structure and texture. Molecular structure may be altered and the rearrangement of chemical components may occur during recrystallisation. Changes in chemical composition due to reactions involving aqueous solutions may also take place. Mineral grains may be physically deformed or broken. The identity of the resultant metamorphic rock depends not only on the chemical composition of the parent rock and that of any incident fluids but also on the temperature and pressure at which the metamorphism took place.

Metamorphism is a process which typically occurs at the high temperatures and pressures present deep within earth's crust. The delineation between sedimentary diagenesis and metamorphism is defined to occur at about 200° Celsius. Changes in mineral form which occur at temperatures lower than 200° Celcius are thus defined as sedimentary diagenesis; changes which take place at higher temperatures are considered to be metamorphic. Metamorphism is defined to occur in the solid state, so that melting of the rock may not take place. The high temperature limit of metamorphism is therefore given by the melting temperature of the rock in question. Any process in which temperatures are great enough that complete melting of the host rock takes place is considered to be igneous rather than metamorphic.

Metamorphic processes may be divided into two categories according to the area of their extent. Regional metamorphism may affect vast expanses of rock, whereas local metamorphism will affect only relatively bounded areas.

Different types of metamorphism may alternately be categorised according to the agents active during each process. Thermal metamorphism is due merely to the influx of heat from a nearby igneous intrusion. In contrast, highly reactive volatile fluids play a significant role in hydrothermal metamorphism. Stress, pressure, and shock are causative agents of dynamic metamorphism.

A third possible method of classification of metamorphic processes is whether or not previously absent elements are introduced into the reaction process. The process of isochemical metamorphism consists of the recrystallisation of previously present elements into new species of minerals. Such minerals may differ structurally from the original minerals, but will contain only those elements which were originally present.

In contrast, allochemical metamorphism integrates elements introduced into the system during the reaction process into the newly formed mineral species. It is apparent that thermal and dynamic metamorphism are types of isochemical metamorphism because they introduce no new elements into the reacting system, while hydrothermal metamorphism is a variant of allochemical metamorphism and may radically change the chemical composition of the system.

Regional Metamorphism

Regional metamorphism describes geologic processes which act over wide areas and cause metamorphic changes in vast expanses of rock. This type of metamorphism is also called dynamothermal metamorphism. Mountain ranges may be composed of large bodies of rock which were shaped and formed by regional metamorphism.

Regional metamorphism occurs in subduction zones where one crustal plate slides under another. At plate boundaries and in places of collision, large quantities of rock may be subsumed and subjected to the greater temperatures and pressures found deep within the earth. This type of regional metamorphic process is called orogenic metamorphism.

Regional metamorphism often occurs beneath midoceanic ridges and trenches. As one oceanic plate slides beneath another or as two plates draw apart, seawater may seep into crust rocks. Both seawater and water escaping from upwelling magmas may catalyse metamorphic processes. Burial metamorphism occurs in large basins which progressively fill with sedimentary material. The rock underneath the basin may sink as the basin fills, experiencing increasing pressures as subsequent levels of sediment are deposited over the top. Metamorphism typically begins at depths of about 8 kilometres where temperatures may be from 100° to 200° Celsius.

Local Metamorphism

Local metamorphism occurs in an area of much smaller scale than that of the great bodies of rock affected by regional metamorphism. Local metamorphism occurs when an igneous body or pluton intrudes into and affects the surrounding rock. For this reason local metamorphism is also called contact metamorphism.

Thermal Metamorphism

The process of thermal metamorphism, also called pyrometamorphism that takes place when an intrusive igneous body heats the surrounding rock. Metamorphism then occurs as heat inflow from the nearby igneous source catalyses recrystallisation of previously present rock.

In cases of thermal metamorphism no new elements are introduced into the set of available elements. Instead new minerals form from those elements which were previously part of the system. For this reason thermal metamorphism is a type of isochemical metamorphism. For example, when a pure limestone ($CaCO_3$) is heated by an intruding igneous body, it will recrystallise into pure marble ($CaCO_3$). However, if impurities such as quartz (SiO_2), clay (kaolinite, $Al_4(Si_4O_{10})(OH)_8$) or iron oxide (haematic, Fe_2O_3, or magnetite Fe_3O_4) are present, new mineral species may be created. Quartz may combine with limestone to form wollastonite ($Ca(SiO_3)$); the aluminium in clay may enter into the reaction and form corundum (Al_2O_3), spinel ($MgAl_2O_4$), and garnet (almandine, $Fe_3Al_2(SiO_4)_3$).

Dynamic Metamorphism

Strain metamorphism occurs when the host rock is subjected to stress; this frequently takes place along fault lines. At low temperatures and pressures the rock may be pulverised; at higher temperatures and pressures ductile flow may occur, and rocks may elongate, fold and flow. Confining non-directional pressures may initiate metamorphism which produces a more dense, compact rock. Shock metamorphism due to extremely high temperatures and pressures of very short duration may occur when initiated by a meteorite impact. Dynamic metamorphism is a type of isochemical metamorphism because no new elements are introduced into the host rock.

Hydrothermal Metamorphism

Hydrothermal metamorphism, also called metasomatism, may take place across wide regions of rock, thereby constituting a variant of regional metamorphism. It may alternately may take place in a limited, localised area and constitute a variant of local metamorphism.

Hydrothermal metamorphism takes place when hot, volatile solutions percolate into and react with the protolith, or the original rock. The heat of the intrusive igneous body and the hot volatile fluids serves to catalyse metamorphic reactions in the host rock. The incident fluids enhance ion mobility in the system and are highly reactive. Both the intrusive magma and the associated volatile fluids may introduce elements which were not present in the protolith into the reaction process. The incident volatile fluids may also dissolve and remove elements originally abundant in the host rock. The chemical composition of the host rock may be extensively altered during the process of hydrothermal metamorphism. Hydrothermal metamorphism is therefore a type of allochemical metamorphism.

The hot, volatile fluids responsible for hydrothermal metamorphism may be derived from several sources. Such fluids may consist of volatiles escaping directly from the intrusive magma; they may also be composed of meteoric groundwater heated by the igneous intrusion or by the geothermal gradient. High temperatures and pressures can expel water from hydrated minerals such as gypsum, talc, chalk, and clay. These sources of fluid are typically associated with local metamorphism. When an oceanic plate is subsumed at a boundary between two plates, water may be transported along with it and may subsequently take part in any metamorphic processes which occur. Seawater may also seep into the crust as two oceanic plates move apart. Such sources of volatile fluids are associated with regional metamorphism and the movement of crustal plates.

Much hydrothermal metamorphism occurs at the boundaries of oceanic plates. Plates which are moving apart allow seawater to percolate through the oceanic crust. As the seawater migrates, it heats and reacts with the host rock. Large quantities of metals such as iron, cobalt, nickel, silver, gold, and copper are dissolved from the crust. When the hot, metal-laden fluid later comes into contact with cold seawater, sulphide and carbonate minerals precipitate to form deposits of metal ores. The copper ore which has been mined on Cyprus for several thousand years is thought to have been deposited in this way.

The presence of previously unavailable elements and the catalysation provided by heat and volatile fluids enables a wide variety of mineral species to form during hydrothermal metamorphism. Many of these species contain elements not originally present in the host rock. The process of hydrothermal metamorphism thus produces a greater variety of minerals than may be formed by thermal metamorphism alone. Silicates which are found in hydrothermal metamorphic deposits include quartz, garnet, wollastonite, olivine, topaz, and tourmaline. The halide fluorite may be present. Sulphide ores such as pyrite, chalcopyrite, sphalerite, and molybdenite and oxides such as magnetite, haematic, spinel, and corundum may also occur.

Local hydrothermal metamorphic deposits may grade into hydrothermal vein deposits. A vein forms when hot, mineral-bearing fluids deposit new materials along the walls of a pre-existing fissure or crack in the host rock. In the case where minerals merely exsolute from the aqueous solution to crystallise along the walls of the fissure, leaving the wall rock intact and unchanged, no metamorphism is involved. However, the minerals of the fissure walls often react with the passing volatile fluids. Some may be dissolved and transported away; some may recrystallise or crystallise as new mineral species. The formation of this type of vein deposit, called a hydrothermal replacement deposit, does constitute hydrothermal metamorphism.

Isochemical and Allochemical Metamorphism

Isochemical metamorphism requires that no new elements are introduced into the system and that the relative concentrations of previously present elements remain constant throughout the metamorphic process. Isochemical metamorphism thus involves no injection or leaching out of new elements by magma or volatile fluids. In contrast, allochemical metamorphism involves changes in chemical composition due to the arrival of new substances or the removal of previously present ones. Thermal metamorphism and dynamic metamorphism are types of isochemical metamorphism, whereas hydrothermal metamorphism typically implies allochemical metamorphism and the introduction of previously absent elements.

Texture

Several different textures are observed in metamorphic rocks.

Foliation

The term foliation indicates a texture sometimes displayed by rocks which have undergone the folding and distortion of dynamic metamorphism. A rock whose mineral grains exhibit a preferred, directional orientation is said to demonstrate foliation. Foliation is occasioned by the contortion of mineral grains during directional and compressional stress. The planar, sheetlike layering of mineral grains in certain minerals is a type of foliation. This layering of grains is termed foliated, lamellar or micaceous habit and is typical of mica and slate. Layered banding where alternating layers of light and dark minerals occur is also a type of foliation.

Porphyroblastic Texture

Certain metamorphic mineral species such as garnet and staurolite tend to recrystallise to form large, individual crystals, while other species such as mica and biotite tend to form masses composed of small interlocked grains. If two species with differing behaviours recrystallise in the same location, the resulting metamorphic rock will typically contain large crystals of one species embedded in a matrix of small crystals of the other. For example, large garnets are often found embedded in a mass of fine-grained muscovite or biotite. Metamorphic rocks possessing this type of texture are termed porphyroblasts.

Banding

Metamorphism sometimes occurs as the minerals of the protolith or host rock become unstable upon the introduction of new elements in solution. The mineral deposits of the host rock morph until equilibrium with the new environmental conditions is reached. The presence of volatile fluids and the influx of heat increase ion mobility and catalyse the metamorphic process. Metamorphism is thus driven by the imperative of chemical equilibrium but regulated by kinetic processes such as the diffusion of ions.

Since metamorphism is regulated by kinetic processes, the innermost portion of a mineral sample may contain the original reactant while the outermost layers are of the new composition and are in equilibrium with the surrounding system. The concentric rings and banded structures which characterise structures of concretionary habit may form as new phases of material surround the original reactant. The vividly coloured concentric layers observed in malachite offer an example of such banded concretions in metamorphic rocks.

Migmatism

Different varieties of rock melt at differing temperatures. For example, the light-coloured, non-ferromagnesian silicate minerals such as quartz, feldspar melt at much lower temperatures than do the dark-coloured, ferromagnesian silicates such as olivene, pyroxene, and amphibole. If a rock deposit reaches temperatures high enough to melt the non-ferromagnesian silicates yet does not reach the melting point of the ferromagnesian silicates, the rock may partially melt although significant portions may remain solid. In some rocks bands of light and dark material will then separate out. The non-ferromagnesian silicates which occupy the light-coloured bands will have melted and will therefore appear to be igneous rocks. The ferromagnesian silicates which occupy the dark-coloured bands, in contrast, will have remained solid and will constitute metamorphic materials. Rocks which contain bands of contrasting colour and disseparate history are termed migmatites. Such formations may be classified as intermediate between metamorphic and igneous rocks.

VEINS

Veins are mineral deposits which form when a pre-existing fracture or fissure within a host rock is filled with new mineral material. The deposition of minerals is typically performed by circulating aqueous solutions. Many ore deposits of economic importance occur in veins.

Vein deposits are believed to form when aqueous solutions carrying various elements migrate through fissures in rock and deposit their burden onto the fissure walls. Hot, rising water escaping from cooling igneous plutons may deposit minerals as it ascends through the crust. As heated magmatic waters rise, the temperature and pressure of their environment drop and minerals exsolute and crystallise. Meteoric ground water may also percolate down through the earth's crust, dissolving surface minerals and gaining heat from the geothermal gradient or from nearby igneous intrusions. At greater depths the dissolved substances may precipitate and crystallise along the walls of the fissures and cavities through which the water travels.

Most vein deposits are formed as new mineral species are precipitated onto rock walls which themselves remain unaltered. In such cases mineral deposits fill the original crack or fissure in the host rock but do not extend into the host rock itself. The boundary between host rock wall and deposited vein minerals therefore remains clearly delineated. Vein deposits of this nature are a type of hydrothermal deposit because the mineral species which compose the veins were precipitated by hot waters. However, sometimes the pre-existing rock wall which contains the vein undergoes alteration. Portions of the host rock may either dissolve and be transported away or else react chemically with the circulating volatile fluids or the newly formed mineral species. In this case the boundary between vein deposit and original rock wall will be unclear. If most of the mineralisation process occurs within the space once occupied by unaltered wall rock then the vein is termed a hydrothermal replacement deposit. A hydrothermal replacement deposit occurs when hot circulating aqueous solutions replace the original rock with new mineral species. This typically occurs in more soluble rocks such as limestone. Hydrothermal replacement deposits are a form of hydrothermal metamorphism or metasomatism.

Hydrothermal Deposits

Hydrothermal deposits are categorised according to the depth and temperature at which they formed. Hypothermal deposits are formed at great depths and high temperatures; mesothermal deposits at intermediate depths and temperatures; and epithermal deposits at the shallowest depths and relatively low temperatures.

Some mineral species crystallise mainly at preferred temperatures and pressures. Because the temperatures and pressures are different for each type of hydrothermal deposit, each has a different, characteristic set of associated minerals.

Hypothermal

Hypothermal deposits are formed at great depths and high pressures and temperatures. Temperatures may range from 300° to 500° Celsius during the formation of such deposits. Casseterite, wolframite and molybdenum veins; gold-quartz veins; copper-tourmaline veins; and lead-tourmaline veins provide examples of mineral associations which may occur in hypothermal deposits. Minerals which are found in hypothermal veins include quartz, fluorite, tourmaline, and topaz. Ore minerals found may include native gold (Au); the sulphides galena (PbS), chalcopyrite ($CuFeS_2$), pyrite (FeS_2), molybdenite (MoS_2), bismuthinite (Bi_2S_3), and arsenopyrite ($FeAsS$); the oxides uraninite (UO_2), cassiterite (SnO_2), and magnetite (Fe_3O_4); and the tungstates wolframite ($(Fe,Mn)WO_4$) and scheelite ($CaWO_4$). Metals which may be extracted from hypothermal deposits consist of copper (Cu), molybdenum (Mo), tin (Sn), tungsten (W), gold (Au), and lead (Pb).

Mesothermal

Mesothermal deposits form at intermediate depths, temperatures, and pressures. Temperatures may range from 200° to 300° Celsius during the formation of such deposits. Quartz and carbonate minerals such as calcite ($CaCO_3$), ankerite ($CaFe(CO_3)_2$), siderite ($FeCO_3$), dolomite ($CaMg(CO_3)_2$), and rhodocrosite ($MnCO_3$) occur in mesothermal deposits. Ore minerals which may be found include native gold (Au) and the sulphides galena (PbS), sphalerite (ZnS), chalcopyrite ($CuFeS_2$), pyrite (FeS_2), bornite (Cu_5FeS_4), arsenopyrite (FeAsS), and tetrahedrite ($(Cu,Ag)_{12}Sb_4S_{13}$). Metals which are mined consist of copper (Cu), zinc (Zn), silver (Ag), gold (Au), and lead (Pb).

Epithermal

Epithermal deposits form at shallow depths under relatively low temperatures and pressures. Temperatures during formation may range from 50° to 200° Celsius. Minerals found include quartz, opal, and chalcedony (SiO_2); calcite ($CaCO_3$), aragonite ($CaCO_3$), and dolomite ($CaMg(CO_3)_2$); the halides fluorite (CaF_2) and chlorargyrite (AgCl); the sulphate barite ($BaSO_4$); native gold (Au); and the sulphides realgar (AsS), cinnabar (HgS), acanthite (Ag_2S), pyrite (FeS_2), orpiment (As_2S_3), stibnite (Sb_2S_3), proustite (Ag_3AsS_3), and pyrargyrite (Ag_3SbS_3). Metals which are mined from epithermal deposits include silver (Ag), gold (Au), and mercury (Hg).

Secondary Replacement Deposits

When exposed to meteoric ground water containing dissolved oxygen and carbon dioxide, primary ore minerals near the surfaces of vein deposits may be altered to secondary minerals. Primary sulphide minerals such as pyrite, chalcopyrite, galena, and sphalerite are in particular vulnerable to oxidisation and alteration. Pyrite (FeS_2) oxidises to goethite ($FeO(OH)$) or melanterite ($FeSO_4.7H_2O$), galena (PbS) oxidises to anglesite ($PbSO_4$) and cerussite ($PbCO_3$), and sphalerite (ZnS) alters to hemimorphite ($Zn_4(Si_2O_7)(OH)_2.(H_2O)$) and smithsonite ($ZnCO_3$). If the presence of carbon dioxide is sufficient, then copper sulphide minerals may alter to the copper carbonates azurite ($Cu_3(CO_3)_2(OH)_2$) and malachite ($Cu_2CO_3(OH)_2$) and the oxide cuprite (Cu_2O) as well as native copper (Cu). Just as primary sulphides alter to secondary sulphates, arsenides or phosphides may oxidise to produce secondary arsenates or phosphates.

Oxidised pyrite and copper sulphides produce sulphuric acid (H_2SO_4) and ferric sulphate ($FeSO_4$) which in turn react with copper to form soluble copper sulphate ($CuSO_4$). Copper sulphate may be transported downwards as meteoric groundwaters percolate through the rock.

As the copper sulphate moves downwards it reacts with the sulphide minerals which it encounters. These are altered to produce secondary copper sulphides such as chalcocite (Cu_2S). The secondary copper sulphides may then in turn oxidise, continuing the downward migration of metal through the vein environment.

The upper zone of a vein in which secondary replacement occurs may become barren and leached and is referred to as the oxidised zone. Delicate crystalline materials may be found in the oxidised zone because the primary minerals have been removed, providing the room necessary for extensive crystals to form. The zone of oxidation is typically fairly shallow; it tends to extend from the surface to the water table. Near the water table a zone of secondary enrichment may form; in this region copper which seeped down from the higher primary regions will be concentrated. Large chalcocite 'blankets' are occasionally found in zones of secondary enrichment. Below the secondary regions only primary minerals are found.

Sedimentary Rocks

Sediments include all uncemented, particulate materials which accumulate at the earth's surface. Such materials are derived from the erosion of pre-existing rocks by mechanical or chemical weathering processes, by precipitation from aqueous solution, or from the skeletal remains of organisms. Sediments are transported by the movement of water, wind, and glaciers. Sedimentary rock is formed when volumes of loose sediments are collected together, compacted, and cemented into rock with the aid of heat, pressure, and cementing agents.

Only five per cent of the earth's crust by volume is estimated to consist of sedimentary rocks. However, at least two thirds of the surface area of the earth is covered by a thin layer of this type of rock. The processes by which sedimentary rocks are formed are surface phenomena driven by surface waters and weather.

Sedimentary Diagenesis

The term diagenesis is used to portray all changes which result in the conversion of loose sediment into sedimentary rock. These changes include recrystallisation, lithification and precipitation. Sedimentary diagenesis takes place at depths of less than a few kilometres within the earth's crust and occurs at temperatures of less than 200° Celsius. Changes in rock which occur at greater depths or temperatures are defined to be metamorphic rather than sedimentary.

Recrystallisation is the means by which a less stable mineral species may take on the form of a chemically identical but more stable mineral species. In order to do this a mineral will dismantle its crystalline lattice and rearrange its chemical constituents. The chemical composition of the system does not change during recrystallisation although the crystalline structure does change. One example of this process is provided by the two calcium carbonate species calcite and aragonite, both of which possess the chemical formula $CaCO_3$. The shells of marine organisms are composed of aragonite; volumes of these shells may accumulate to produce large deposits of sediment. Because aragonite is less stable than calcite, when these deposits are buried and thereby subjected to increased temperatures and pressures the aragonite recrystallises to adopt the structure of calcite. Calcite remains stable in more extreme conditions than does aragonite, and is the main constituent of the familiar sedimentary rock called limestone.

The methods by which unconsolidated particles of sediment are transformed into sedimentary rock are termed lithification. Lithification includes the two processes of compaction and cementation. Compaction is a physical change which occurs as layers of sediment accumulate in one area. Gradually the lowest layers are compressed by weight of the layers above; the amount of pore space between individual grains of sediment is reduced, and water trapped within the pore spaces is forced out. Cementation, in contrast, is a chemical change. During cementation elements carried by aqueous solutions which seep through the sediment precipitate into the pore spaces between grains, filling these spaces and cementing the individual particles together. The most prevalent cementing minerals are calcite ($CaCO_3$), silica (SiO_2), and iron oxide (Fe_2O_3). The presence of iron, which is a strong pigmenting agent, tints rocks cemented with iron oxide orange, red, or red-brown.

Not all sedimentary rocks undergo lithification; some are crystalline and originate as masses of interlocked crystals. Certain limestone deposits, which form when deposits of aragonite recrystallise into calcite, offer examples of crystalline sedimentary rock.

Detrital and Chemical Sedimentary Rocks

Sediments and the particulate mineral matter which compose sedimentary rocks are derived by the two different means of mechanical and chemical weathering. Sedimentary rocks may in turn be classified according to which of these two means resulted in their formation. Detrital sedimentary rock contains particles and sediment which were derived by mechanical means, while chemical sedimentary rock contains only substances which were derived by chemical weathering processes. The materials which compose detrital sedimentary rocks were ground down by mechanical processes such as wearing by wind, grinding by glaciers, and splitting due to cyclic freezing and thawing.

These materials were also transported to the site of deposition by mechanical means such as wind, suspension in moving water, or travelling glaciers. These are compacted and cemented together; few chemical changes occur in the mineral material of which the rocks are composed. The constituent materials of chemical sedimentary rock, in contrast, were dissolved in water, transported to the deposition site in the form of aqueous solutions, and finally precipitated.

Attack by rainwater and atmospheric carbon dioxide (CO_2) may dissolve parent rock and allow it to be transported to the site of deposition. The precipitation of chemical sedimentary rocks may then be caused by either inorganic or organic agents. Inorganic processes include evaporation whereas organic processes occur when water-dwelling organisms extract calcite from their environment and in turn excrete a calcareous exoskeleton.

Sandstone provides an example of a familiar detrital sedimentary rock while limestone is perhaps the most well-known chemical sedimentary rock. Sandstone is so named because it is composed of sand-sized grains. Because the mineral quartz (SiO_2) is quite stable and durable, it forms the most prominent constituent of most sandstones. Less stable or robust minerals often disintegrate during the mechanical weathering and collection processes which result in detrital sedimentary rocks.

Limestone is an example of a chemical sedimentary rock which may be formed through either inorganic or organic processes. Limestone stalactites ($CaCO_3$) offer an example of inorganic precipitation. Such stalactites are formed as calcium-carbonate bearing water drips from an overhead projection in a cave. (The Greek word stalaktos means 'dripping' or 'trickling'.) A minute quantity of calcium-carbonate is precipitated from each water droplet, slowly building up the pendulous stalactite formation. Many larger limestone deposits, on the other hand, are formed through organic precipitation. In particular this occurs when small organisms such as corals extract calcite from their ocean environment and secrete a calcareous extoskeleton. A colony of corals may over time achieve a massive, solid limestone reef.

Chemical sedimentary minerals which precipitate when evaporation removes the water in which they are dissolved are termed evaporites. These minerals include halite (NaCl), sylvite (KCl) and gypsum ($CaSO_4 . 2H_2O$).

Texture

The quality of texture also plays a role in the classification of sedimentary rocks. With respect to texture, sedimentary rocks are categorised according to whether they are clastic or non-clastic. The word clastic is derived from the Greek term klastos, which means 'broken'. Rocks of clastic texture exhibit a mass of discrete particles and grains which are cemented together but retain their individual shape. All detrital sedimentary rocks display clastic texture, while some chemical sedimentary rocks display clastic texture and others display non-clastic texture. Chemical sedimentary rocks such as halite (NaCl) and sylvite (KCl) which are deposited when seawater evaporates possess a non-clastic texture. Rocks such as calcite, which is formed as aragonite recrystallises, typically have a non-clastic texture composed of interlocking crystals.

Description of Specimens

My small collection of minerals contains five species, all of which were removed from the copper mines of Butte, Montanna during the 1950's. Of these five species, native copper, the copper carbonates azurite and malachite, and the amorphous copper silicate chrysocolla are copper-bearing substances. Rhodocrosite ($MnCO_2$) is, like azurite and malachite, a carbonate mineral. However, rhodocrosite is an ore of manganese rather than copper.

Copper (Cu) is a native metal which has long been utilised by humans as a natural resource. The crystalline lattice of this mineral contains only elemental copper and its structure is derived from metallic bonds. Like all metals, native copper is dense, soft, malleable, and ductile. It conducts well. It is opaque and demonstrates the shiny, highly reflective lustre typical of metals. In colour it is the familiar, rosy copper red; it possesses a red streak. Exposed surfaces may tarnish to black oxide or green carbonate.

Native copper possesses a face-centred cubic lattice, which is the most dense and symmetric of all possible crystalline structures. It is of the isometric or cubic crystal system. Individual copper crystals may be of cubic, dodecahedral, tetrahedral, and very rarely of octahedral shape. Aggregates of crystals may be wiry in habit or form a dendritic, arborescent branching structure. Mineral specimens demonstrating whole individual crystals and well-formed crystal faces are relatively rare.

The two minerals azurite ($Cu_3(CO_3)_2(OH)_2$) and malachite ($Cu_2Co_3(OH)_2$) are closely related. Both are hydrous copper carbonates, which contain the carbonate anion group, $(CO_3)^2$, the hydroxyl anion (OH^-), and the copper cation (Cu^{2+}). Azurite is royal blue or brilliant blue in colour while malachite may be bright to dark green. The colouration in both species is idiochromatic and is due to the presence of copper, which is a chromophore and possesses a strong pigmenting effect. Malachite's green results from the fact that the copper which it contains is more highly oxidised than that of azurite.

Aggregates of crystals of both species are frequently botryoidal or mammillary in habit. Large individual crystals of malachite are rare. More commonly the mineral occurs in rings, bands, and concentric layers composed of varying shades of green. Both azurite and malachite occur as earthy, incrusted masses. Rhodocrosite ($MnCO_3$) is member of the carbonate class and an ore of manganese. Its characteristic rose pink or red colour is idiochromatic and is due to the presence of the chromophore manganese (Mn). Rhodocrosite is translucent and will transmit light diffusely; it possesses a vitreous or pearly lustre. Like azurite and malachite, rhodocrosite is a member of the carbonate class. However, rhodocrosite contains neither water (H_2O) nor the hydroxyl anion (OH^-) and is therefore not a hydrous carbonate.

Instead it is a member of the calcite group. Rhodocrosite is like all calcite minerals a member of the rhombohedral crystal system. Rhodocrosite's cleavage is perfect in three directions and provides external indication of its rhombohedral internal structure. Crystals are rare but may be rhombohedral in shape; more typically the mineral demonstrates botryoidal, encrusting habit.

In colour, chrysocolla is green to sky-blue. This colouration is idiochromatic and is due to the presence of the chromophore copper (Cu). Chrysocolla is translucent; samples may be of vitreous or glassy lustre or appear greasy, dull, or earthy. It is very brittle, sometimes fragile; it has no cleavage, and demonstrates uneven or conchoidal fracture. Specimens of chrysocolla are typically of massive or earthy habit.

The substance may also display reniform or bulbous, botryoidal habit. It lacks macroscopic crystals. Azurite, malachite and chrysocolla are commonly found in the oxidised zones of copper veins and deposits. Rhodocrosite occurs in veins and hydrothermal replacement deposits in the company of manganese minerals as well as other metal ores such as copper.

Native Copper

Native copper has been utilised by humans as a natural resource since perhaps 3,400 B.C. The word 'copper' is derived from the Latin *cuprum*, which was in turn a modification of the Greek name *Kyprios*. The word *Kyprios* indicated the island of Cyprus, where the metal was mined in ancient times.

Copper is a member of the class of native elements. The minerals of this class either contain only the atoms of a single element or else are metal alloys. A metal alloy may contain two or more metallic elements in solid solution, but its constituent atoms must be bonded only by metallic bonds. Native copper is a metal and its lattice structure contains only elemental copper.

Native copper gains its structure and many of its properties from metallic bonds. When metal atoms bond together into a crystal the valence electrons become delocalised and are free to roam throughout the lattice. The attraction of the positively charged atoms for the surrounding electrons bonds the atoms into an ordered structure. The positive ions surrounded by the negatively charged electron reservoir retain an almost perfectly spherical shape.

In a pure metal, these atoms are all of one element and are therefore all of the same atomic radius. Because the atoms are spherical, equally sized, equally charged, and surrounded by negative, delocalised electrons they pack closely together to form a dense and very ordered crystalline structure. Metals tend to be dense and to have a high degree of symmetry. Copper possesses a typically high density of 8.9 g/cm^3.

Parallel planes of equally sized and charged atoms may glide across one another, resulting in the softness, malleability, ductility, and sectility which are characteristic of metals and are demonstrated by native copper. As expected, copper is soft, possessing a hardness of 2½ to 3 on the Mohs scale.

The many delocalised valence electrons present within a metallic lattice serve as conducting electrons and account for the high conductivity of metals. Copper, like all native metals, conducts well.

Electromagnetic waves cannot propagate through a metal.

Instead, they are absorbed and reflected. The absorption index of a metal is very high, and nearly all incident light is reflected. This renders the metal completely opaque and makes it appear shiny and brilliant. Copper is therefore opaque and will not transmit light in thin section; the mineral possesses a typical shiny, reflective metallic lustre.

In colour, native copper consistently demonstrates the familiar and distinctive copper red; its streak is also red in hue. The red of native copper is an example of idiochromatic colouration, or colour which is derived directly from the presence of one of the main constituent elements of the mineral. Idiochromatic colour is a property which is directly related to the chemical composition of a mineral species and is therefore possessed by all specimens of that species. This characteristic red of native copper is attributable to the presence of copper, which has strong pigmenting capabilities and is the only constituent of the crystal lattice. Exposed surfaces of native copper crystals may tarnish to black oxide or green carbonate according to the following reaction:

$$2Cu + O_2 \rightarrow 2CuO$$

$$2Cu + 2CO_2 + O_2 \rightarrow 2CuCO_3$$

Native copper has no cleavage and does not tend to break along regular planes. Its fracture is 'hackly', or jagged, containing many small irregularities.

The lattice structure occupied by copper is that of cubic closest packing or face-centred cubic. This type of structure is based around a cube on which an atom occupies each of the eight corners. Atoms are emplaced at the centre of each of the six exterior faces as well. This is the most dense of all possible lattice structures. Native copper is of the isometric crystal system. Isometric crystals possess four three-fold axes of symmetry and are measured against three axes of equal length which occur at right angles to one another.

Individual copper crystals may be of cubic, dodecahedral, tetrahedral, and very rarely of octahedral shape. Aggregates of crystals may be wiry in habit or form a dendritic, arborescent branching structure. Fine mineral specimens with whole individual crystals and well-formed crystal faces are quite rare.

Azurite and Malachite

The mineral azurite ($Cu_3(CO_3)_2(OH)_2$) is an ore of copper, and provided human civilization with one of its earliest sources of this metal. Both azurite and its close relative malachite have been mined for thousands of years. Because of its amazing blue colour, azurite has long been used as a pigment in dyes and as a decorative gemstone. The species was named 'azurite' in honour of this distinctive azure blue. The mineral is also called 'chessylite', a name reminiscent of Chessy, France, where large quantities have been mined.

Single crystals of azurite are a dark, royal blue; the mineral exhibits a brilliant azure blue colour when it is of dull or earthy lustre. Its streak is blue. This colouration is idiochromatic and is due to the presence of the chromophore copper, which grants a strong pigmenting effect, within the crystal lattice. Azurite's characteristic blue colour is useful in identification.

Azurite produces some specimens of adamantine lustre which are very sparkly and highly translucent with a high refractive index. It forms other samples of vitreous or glassy lustre and medium refractive index; its surface may also be rough, giving a dull, earthy and almost lustreless appearance.

Azurite possesses a density of 3.77 to 3.83 g/cm^3 and a Mohs scale hardness of $3^1/_2$ to 4. Its cleavage is good in two directions; its fracture is conchoidal, or shell-like, resulting in a series of concentric rings about the stressed point. The mineral is of brittle tenacity and is not malleable. Like other minerals of the carbonate class, azurite effervesces slightly in hydrochloric acid. This provides a useful means of identification in the field.

In shape azurite crystals may be equidimensional; prismatic, or elongated in one dimension; or tabular, appearing in flat plates. Aggregates of crystals are frequently botryoidal in habit, having a globular form said to resemble a bunch of grapes. Azurite may also form radiating crusts or earthy masses.

MALACHITE

The mineral malachite is a semi-precious stone and a common ore of copper. Its name is derived from the Greek term malache, 'mallow', which refers to its leaf green colour. Like azurite, malachite has been used for centuries as a pigment in dyes and as a semi-precious stone.

Malachite may vary from bright to dark green in colour; its streak is consistently light green. The colouration of malachite is idiochromatic and due to the presence of copper within the chemical formula ($Cu_2Co_3(OH)_2$). Specimens may be of adamantine, silky, or dull lustre.

Malachite possesses a density between 3.6 and 4.0 g/cm^3 and a hardness of 3½ to 4 on the Mohs scale. It demonstrates perfect cleavage in one direction; however, this cleavage occurs rarely. More frequently observed is the conchoidal or splintery fracture.

Large individual crystals of malachite are very rare, but may occur in prismatic form. Aggregates of crystals are frequently mammillary, forming smooth, bulbous masses. They may also be botryoidal and globular. They may occur in radial assemblages, forming layers and concentric rings in varying colours of green. Malachite is likewise observed to produce fibrous threads and tufts, stalactitic formations, and earthy, incrusted masses. Distinctive concretionary banding patterns occur in many samples, particularly in botryoidal masses. These are caused by varying oxidation levels in the aqueous solution from which dissolved minerals exsolute. The mechanism by which this occurs is not fully understood.

Classification and Association of Azurite and Malachite

Both azurite and malachite are hydroxide-containing members of the copper carbonate class. Carbonate minerals compose a class in which $(CO_3)^{2-}$ anions are linked by various cations within the unit cell. Copper carbonates in particular are hydrous carbonate minerals which contain both the copper cation (Cu^{2+}) and hydroxyl anions (OH^-). Azurite possesses the chemical formula $Cu_3(CO_3)_2(OH)_2$; the formula of malachite is $Cu_2(CO_3)(OH)_2$. Azurite and malachite are commonly found together and in conjunction with native copper and other copper ores.

Crystal System and Molecular Structure

Both azurite and malachite belong to the monoclinic crystal system. Members of this crystal system possess three axes of unequal length, two of which are perpendicular to each other. The end faces are inclined relative to the side faces rather than being orthogonal to them. Such crystals are usually short, having the appearance of a distorted rectangle.

The molecular structure of azurite consists of square groups composed of two (O^{2-}) and two (OH^-) anions attached to a single, central Cu^{2+} cation. The anions may be paired along one side of the square or across its diagonal. These square groups are then linked in chains by plane triangular $(CO_3)^{2-}$ anions.

In the molecular scheme of malachite a single Cu^{2+} cation is surrounded by an octahedron composed of O^{2-} and (OH^-) anions. This octahedron is composed of four of either anion (O^{2-} or (OH^-)) and two of the other. Many such octahedra in turn form chains linked by $(CO_3)^{2-}$ groups.

Colour

Azurite and malachite have very similar chemical formulae:

Azurite: $Cu_3(CO_3)_2(OH)_2$ or $Cu(OH)_2 . 2(CuCO_3)$

Malachite: $Cu_2Co_3(OH)_2$ or $Cu(OH)_2 . (CuCO_3)$

The vibrant greens and blues exhibited by these two minerals are due to the inclusion of copper, which is a very effective pigmenting agent, among their chemical constituents. This type of colouration is termed idiochromatic.

Malachite possesses a ratio of 1:1 $Cu(OH)_2$ to $CuCO_3$ while azurite possesses a ratio of 1:2 for the same substances. Since $Cu(OH)_2$ is more highly oxidised than $CuCO_3$ and malachite possesses the higher ratio of this cation, malachite thus occupies a later stage in the oxidation process than does azurite. The more prevalent oxidation of malachite is responsible for its bright green colour as compared to the deep blue of azurite.

Pseudomorphism in Azurite and Malachite

Malachite is more stable than azurite; over time azurite will invert to malachite. Sometimes the external form of the original azurite is preserved during this process. Malachite can thus form a pseudomorph of azurite as azurite unit cells are replaced over time by those of malachite. When a pseudomorph of a crystal forms, the unit cells of the original mineral are gradually replaced cell for cell by those of a new mineral. The original macroscopic crystal shape remains intact, but the chemical substance of which it is composed is now that of a different mineral. Although the external, macroscopic form has been preserved, the internal microscopic structure has changed. A pseudomorph thus possesses the chemical composition of the new mineral species while its external crystal form preserves the shape of the original, replaced species.

A pseudomorph of malachite after azurite retains the same shape as the original azurite crystal but is composed of malachite rather than azurite. The pseudomorph is therefore malachite green in colour rather than azurite blue. The chemical formula describing the inversion of azurite to malachite is:

$$2\,[Cu(OH)_2.\,2(CuCO_3)] + H_2O \rightarrow 3[Cu(OH)_2.(Cu\,CO_3)] + CO_2$$

2 azuriite + water → 3 malachite + carbon dioxide

Mineral specimens containing only azurite, only malachite, and varying portions of each substance exist. The contrast between azurite's intense blue and malachite's bright green is very pleasing to the eye. Samples in which the transformation process has begun but remains incomplete can therefore be quite beautiful.

Rhodocrosite

Rhodocrosite ($MnCO_3$) is member of the carbonate class and an ore of manganese. Its name is derived from the Greek words *rhodon,* meaning 'rose', and *chros,* colour. The name was given in reference to the mineral's characteristic rose pink or red colour. Rhodocrosite's distinctive pink colour is an example of idiochromatism which is due to the presence of the chromophore manganese (Mn) in the chemical formula. This characteristic pink colour may darken upon exposure to the atmosphere. Rhodocrosite is known to alter to black manganese oxides or hydroxides. The mineral displays a white streak. It is translucent and will transmit light diffusely; it possesses a vitreous or pearly lustre. Rhodocrosite has a density of 3.4 to 3.6 g/cm^3 and a hardness of $3^1/_2$ to 4 on the Mohs scale. It is brittle and displays uneven fracture.

Like the hydrous copper carbonates azurite and malachite, rhodocrosite is a member of the carbonate class. The carbonate minerals are compounds of a metal or semi-metal with the carbonate anion $(CO_3)^{2-}$. Rhodocrosite is not, however, a hydrous carbonate because it contains neither water (H_2O) nor the hydroxyl anion (OH^-). Instead it is a member of calcite group. The minerals of the calcite group possess a simple geometry in which layers of $(CO_3)^{2-}$ radicals alternate with layers of (2+) metallic cations. Ionic bonds join the carbonate anions to the metallic cations.

Rhodocrosite and all calcite group minerals are of rhombohedral crystal system. Crystals of this system are measured against three vertical axes which intersect one another at 120° angles and one horizontal axis which is perpendicular to the other three. A rhombohedron may be envisioned as an equilateral parallelogram which has been turned up onto one corner. The cleavage of rhodocrosite is perfect in three directions. Like the cleavage of all minerals of the calcite group, it is rhombohedral. Crystals are rare but may be rhombohedral; more typically the mineral demonstrates botryoidal, encrusting habit.

Rhodocrosite is found in veins and hydrothermal replacement deposits in the company of manganese minerals as well as other metal ores. It may alter to manganese oxides or hydroxides.

CHRYSOCOLLA

The hydrous silicate chrysocolla ($Cu_2H_2Si_2O_5(OH)_4$) is an ore of copper. The name of the species is derived from the Greek words *chrysos*, meaning gold, and *kolla*, 'glue'. It was given this name because it strongly resembled a similar material which was once used to solder gold. Chrysocolla is a hydrous or basic copper silicate because it contains the hydroxyl anion (OH^-). Unlike true minerals it does not possess a crystalline lattice. Instead it is an amorphous 'silica gel' or gelatinous precipitate. Because it is amorphous and lacks a crystalline structure, chrysocolla is not a mineral in the strictest sense. However, it is a copper-bearing solid substance which is found in the oxidised zones of copper veins and is associated with minerals such as azurite, malachite, cuprite, and native copper.

In colour chrysocolla is green to sky-blue. This colouration is idiochromatic and is due to the presence of the chromophore copper (Cu) within the material. The streak is very pale blue or green. Chrysocolla is translucent and is able to transmit light when sliced into thin sheets. Samples may be of vitreous or glassy lustre; they may also appear greasy, dull, or earthy.

Chrysocolla is quite soft, possessing a hardness of 2 - 4 on Mohs scale; it has a density of 2.0 - 2.4 g/cm^3. It is very brittle, sometimes fragile. It has no cleavage and demonstrates the uneven or conchoidal fracture which is typical of glassy, amorphous materials. This type of fracture displays a series of rounded, concentric rings radiating away from the point of impact. Conchoidal fracture may sometimes be observed on a piece of thick broken glass such as the base of a glass bottle.

Specimens of chrysocolla are typically of massive or earthy habit. The substance may also display reniform or bulbous, lumpy botryoidal habit. It lacks macroscopic crystals. An amusing field test may aid in the identification of chrysocolla. When touched to the tongue a specimen of this substance will usually adhere lightly. It sometimes tastes bitter or basic.

7

Classification of Minerals

The basic division of all minerals into classes, according to the types of chemical compounds, with slight variations, has become fundamental in the modern classification of minerals. But the sub-division of classes into sub-classes and groups of minerals has changed essentially, especially in the last 30-35 years, due to outstanding advances in the study of crystalline substances.

X-ray structural analysis of crystals on the basis of Y.S. Fyodorov's regular arrangements of points in space has proved the existence of an interrelation between chemical composition and crystal structure, and this has led to the more accurate determination of the chemical formulas of many complex compounds. In this way it has been established that such compounds as the minerals of the spinel group, titanates, and columbates-tantalates heretofore regarded as separate classes, are actually complex oxides. Similarly our concepts of the chemical nature of silicates underwent profound change when complex anions of most diverse form and structure were found within their crystal structures.

These advances have led to the discovery of a functional dependence of not only the external shape of the crystals, but also of their optical, electrical, and mechanical (hardness, cleavage, elasticity) properties on the specific features of their crystal structure.

These properties of minerals proved to be functionally related both to the spatial arrangement of the structural units (ions, atoms, molecules) at a definite ratio between their sizes, and on the type of bonding between them, as well as the properties of the ions and atoms themselves determined by their position in the periodic table and, therefore, by their structure.

Furthermore, a knowledge of the dependences governing the combinations of chemical elements in the course of the formation of minerals has deepened our understanding of mineral genesis. This knowledge also helped to understand the sequence and order of the "selection" and combination of particular ions in the course of the formation of crystals segregating from a solution or melt, depending on the physicochemical factors of the equilibrium of the system (temperature, pressure, and the concentration of the chemical components).

In this way it is also possible to account for the regularities in the distribution of the chemical elements and types of chemical compounds in various successive magmatic products (igneous rocks, pegmatites, hydrothermal formations) or in the products of exogenous processes which result in the formation of the crust of weathering and of chemical sediments in aqueous basins.

All this indicates that a new page is opening in the history of mineralogy. It is no longer possible to understand the properties of minerals out of context of chemical composition and without analysing their crystal structures. And *vice versa,* even when we do not know yet the crystal structure of a mineral, we may, on the basis of its physical and chemical properties, form an idea of the particular structural type of compound to which it belongs.

Therefore, *the classification of minerals should be crystallochemical and based on the knowledge of the interrelationship of all the properties of natural chemical compounds and their chemical composition and crystal structure.*

All natural formations comprising the subject of mineralogy, as well as artificial substances in chemistry, are divided into two big independent groups:

(i) *Inorganic minerals* which, apart from rarely occurring native elements, include natural compounds (except organic compounds) of all the chemical elements;

(ii) *Organic minerals* comprising diverse compounds of carbon (except carbonates and carbides which are classed with the inorganic compounds).

Organic compounds, as known, differ materially from inorganic compounds not only in chemical properties but also in crystal structure and the type of bonding between the structural units. While the present state of knowledge of inorganic minerals is quite advanced, the mineralogy of organic minerals is almost rudimentary.

Twenty years ago such natural substances as peats, coals, lignites, resins and petroleum were regarded as independent minerals. However, recent investigations, particularly of solid inorganic substances, by means of new techniques established beyond doubt that these substances are nothing but widely heterogeneous mixtures of a number of organic minerals the true nature of which has only now become the object of study.

The classification of *inorganic* minerals is based on the following principles. Since minerals are basically crystalline products of chemical reactions, for our purposes it is best to classify them according to chemical composition and crystal structure. It is these features that are related to the most important chemical and physical properties by which minerals are identified and the conditions of their formation determined.

All the minerals of inorganic nature are first divided by the chemical criterion into large classes which differ in the *type of chemical compound* and in the type of chemical bonding between the structural units. A special position is held by *native elements,* mostly metals, with the characteristic metallic bonding between the atoms. Intermetallic salt-like compounds are also classified with these. An independent division is formed by *sulphides* and similar compounds, which with regard to a number of their properties occupy an intermediate position between typical metals and compounds with ionic bonding. The so-called sulphosalts, about which little is known structurally, tentatively are put into this division.

Next comes a characteristic type of chemical compounds with typically ionic bonding, the *halides,* i.e., compounds of metals with fluorine, chlorine, bromine, and iodine. These are followed by *oxides* and *hydroxides* (simple and complex compounds of metals with oxygen or a hydroxyl). Finally, there is the class comprising numerous *salts of oxygen acids,* i.e., compounds of metallic cations with different oxygen-containing complex anions.

According to the type of anions, some of these divisions must be further divided into *classes* and *sub-classes.* The latter are further sub-divided into *groups* by the token of similarity of crystal structure and specific chemical features which, in the main, depend on the cations entering into their composition. Thus, a classification of inorganic minerals may be presented as follows (without sub-division into groups):

Division I. Native elements and intermetallic compounds.

Division II: Carbides, nitrides, and phosphides.

Division III: Sulphides, sulphosalts, and similar compounds.

Class 1: Simple and binary sulphides and similar compounds.

Class 2: Sulphosalts.

Division IV: Halides.

Class 1: Fluorides.

Class 2: Chlorides, bromides, and iodides.

Division V: Oxides.

Class 1: Simple and complex oxides.

Class 2: Hydroxides.

Division VI: Oxygen Salts.

Class 1: Iodates.

Class 2: Nitrates.

Class 3: Carbonates.

Glass 4: Sulphates and selenates.

Glass 5: Chromates.

Class 6: Molybdates and tungstates.

Class 7: Phosphates, arsenates, and vanadates.

Class 8: Arsenites.

Class 9: Borates.

Class 10: Silicates.

A. Silicates with isolated tetrahedra of the SiO_4 anion.

B. Silicates with isolated groups of SiO_4 tetrahedra.

C. Silicates with continuous chains of SiO_4 tetrahedra.

D. Silicates with continuous sheets of SiO_4 tetrahedra.

E. Silicates with continuous three-dimensional networks of SiO_4 and AlO_4 tetrahedra.

To avoid repetition, the dependence of the main properties of minerals on composition and crystal structure will be given in the introductions to each respective division, class, subclass, and group, inasmuch as it is this dependence that serves as a basis for classifying them. In describing separate minerals, only those additional properties will be given which depend on their particular structure and composition.

MINERAL SPECIES AND VARIETIES.

The basic classification unit of natural chemical compounds is a mineral species generally possessing a definite chemical composition and a definite crystal structure. Thus, polymorphous modifications of one and the same crystal substance constitute separate mineral species (graphite and diamond, a-sulphur and jS-sulphur).

When two isostructural substances form a continuous series of solid solutions (e.g., Au—Ag, $MnWO_4$—$FeWO_4$, $NaSi_3AlO_8$—$CaSi_2Al_2O_8$), physico-chemically we have a single phase. In mineralogical practice, however, it is customary to designate by several names not only the end members of the series but also the intermediate forms of an isomorphous series (for instance, gold—electrum—silver; huebnerite— wolframite—ferberite; albite—oligoclase—andesine—labradorite—bytownite—anorthite).

What is meant by *varieties* are minerals which are either identical or almost identical in crystal structure, and only slightly differ in:

(1) Chemical composition, one of the chemical constituents of the mineral being isomorphously replaced in part by another constituent; e.g., cobalt-pyrite, (CoS_2), a variety of pyrite which may be distinguished from the latter only by chemical analysis;

(2) Certain physical properties, though the minerals are exactly the same in regard to composition and structure; e.g., amethyst differs from colourless quartz by its violet colour; or a sooty variety of pyrolusite (MnO_2) which differs from its crystalline individuals (polyanite) only in the degree of dispersion and in false low hardness;

(3) Both in composition and physical properties, e.g., a strongly ferruginous variety of sphalerite (ZnS) differs from the common light-coloured species by being almost black, and in certain other features.

Each mineral species of a definite chemical composition has a name of its own. The majority of polymorphous modifications also have either specific names or the same name prefixed by Greek letters, a, *y* — As to mineral varieties, formerly it had been the custom to give them separate names, even when the difference in composition was negligible. By now many of these names have been abandoned, the only exceptions being gemstones (emerald, ruby, sapphire, amethyst, citrine, morion, etc.). Mineral varieties differing in composition are usually called by the names of their species with either an adjective or a prefix indicating some of their properties (magnesiomagnetite, strontium-aragonite).

Native Elements and Inter-metallic Compounds

Thirty-odd native elements, mostly metals are found in the earth's crust. Also classed with these are a number of gases and very few elements occurring in the liquid state (mercury and some amalgams).

The total weight of the native elements in the earth's crust is very small (less than 0.1 per cent of the total mass). Of this about 0.04 per cent falls to the share of nitrogen and 0.01 to 0.02 per cent to the share

Elements occurring native (bold and medium type).

The remainder of the native elements therefore constitute not more than 0.05 per cent. They are represented by hydrogen, argon, helium, carbon, sulphur, gold, elements of the platinum group, copper, and bismuth.

Some chemical elements are found in nature almost exclusively in the native state, and are called *noble* elements. In the table they are shown bold type.

In the first place these are the noble gases He, Ne, Ar, Kr, Xe, and Rn outstanding for their chemical inertness, i.e., inability to form compounds with oxygen, hydrogen, and other elements. This property of the noble gases is explained by the fact that their atoms have stable outer shells of two or eight electrons.

A special position in the table is held by the group of noble metals (Ru, Rh, Pd, Ag, Os, Ir, Pt, and Au). These are found in the fifth and sixth long periods of the periodic table. In these periods, the atomic radii are almost identical in the vertical groups (the phenomenon known as lanthanidic contraction of atomic volumes). This circumstance plays a very important role in the chemical affinities of these elements in natural conditions, especially in the formation of isomorphous mixtures of noble metals. The metals of the platinum group which form solid solutions are always found together in deposits. The triads of Fe, Go, and Ni in the composition of these minerals usually hold a subordinated place. Isomorphism is even more pronounced in the case of Au and Ag, but Cu rarely occurs in them in solid solution.

Of the other native metals found in the subperiods on the right-hand side of the table, the so-called semi-metals (As, Sb, and Bi) are the most widespread. Despite some similarity in chemical properties, these elements occur in nature in different conditions. Only arsenic and antimony occasionally form the intermetallic compound AsSb.

Of the native non-metals, free hydrogen containing impurities of other gases, is often found in rocks and economic minerals. Nitrogen and oxygen, as known, are the principal constituents of the atmosphere. Carbon usually occurs in two modifications which differ in crystal structure. Native sulphur is for the most part the product of partial oxidation of H_2S and, less often, of the reduction of SO_2, as well as of certain sulphates and sulphur-rich organic compounds.

As many as 80 mineral species and varieties are classed under this division by the criterion of their chemical composition. This greatly exceeds the number of their constituent elements. Some of the elements occur in two or more *polymorphous* modifications (diamond, graphite, a-sulphur, jS-sulphur, etc.). Other elements which occur in the native state, may also form *solid solutions* with each other, e.g., electrum (Au, Ag) and palladic platinum (Pt, Pd).

Also common are *intermetallic* compounds which have stoichiometric formulas and are distinguished by specific features of crystal structure (e.g., algodonite—Cu_3As, stibiopalladinite—Pd_3Sb, discrasite— Ag_3Sb, etc.). Along with compounds of a definite composition there are others the composition of which is variable, e.g. (Pt, Fe), (Pt, Cu), etc. The number of intermetallic compounds that are obtained artificially is truly enormous.

Crystal Structure and the Physical Properties of Minerals

Native elements occurring in the solid state have different crystal structures. As far as the type of atomic linkage is concerned, predominant are the most closely packed structures with metallic bonding, occasionally with features of bonding which is transitional between heteropolar bonds and van der Waals forces. Some essential common properties of native metals depend upon these characteristics.

Of all natural substances, native *metals* are the best conductors of heat and electricity. When polished, all of them possess a strong metallic lustre, i.e., high reflecting power. The refractive indices on which lustre depends are of the highest order. Only in the case of gold, silver and copper are the refractive indices less than 1.0, but high enough to make them highly reflective. The indices of absorption of light of these metals are also exceptionally high.

The overwhelming majority of native metals are silver or tin-white. The striking exceptions are gold and copper. These elements may display valencies higher than one would expect from their position in the periodic table ($CuCl_2$, $AuCl_3$). Probably the selective reflection of light is caused by the specific features of the structure of the nuclei of these elements. Of all known minerals, native metals possess the highest specific gravity (especially in the platinum group of minerals). Those metals whose structure has a coordination number 12 and which have no directive bonds, are highly malleable, have no clearly expressed cleavage and their hardness is low. The only exception as far as hardness is concerned is iridium and minerals containing a high proportion of iridium.

The malleability and plasticity of metals chiefly depend on the number of directions to which the most closely-packed sheets are at right angles. In media with cubic closest packing, there are four such directions, whereas in the hexagonal structure there is only one. Therefore, the metals with structures of cubic packing have greater plasticity than those of hexagonal structure. Owing to the lanthanidic contraction of atomic volumes, noble metals are more inert chemically. This is why gold and metals of the platinum group are widely distributed in valley placers.

The *semi-metallic* group (arsenic, antimony, bismuth), whose cubic structures are, so to speak, partially distorted, differ somewhat from typical metals in physical properties. The smallest distortions in crystal structure are observed in bismuth, which has the strongest lustre and the lowest brittleness and hardness in this group. All the three metals have common morphological features and cleavage directions, which strictly follows from the specific features of their crystal structure.

The metalloids are entirely unlike typical metals in their crystal structures and properties depending on them. They will be characterised in detail in the descriptions of individual minerals.

Gold Group

This group includes typical native metals, such as copper, silver, and gold, and their varieties in terms of chemical composition. Of these elements, gold occurs in nature mostly in the native state and rarely in the form of tellurides. Silver frequently occurs in the form of sulphides and halides. Copper in the native state occurs much more rarely than in compounds with sulphur, oxygen and other elements.

All these metals crystallise in the cubic system, forming crystal structures of the same type. Gold and silver, ,the unit cells of which are almost equal in size, form solid solutions, whereas copper, whose unit cell is smaller, forms solid solutions with gold only at high temperatures and these disintegrate on cooling. For the silver-copper system, the limits of their solubility in one another are extremely narrow.

COPPER, Cu: Chemical Composition: Usually chemically pure, sometimes contains impurities: Fe (up to 2.5 per cent), Ag (often as inclusions of native silver), Au, rarely as a 2-3 per cent solid solution *(auric copper).*

Crystal structure of the simplest type is a face-centred cube with the closest packing of atoms. Cu atoms are distributed at the apexes of the cube and at the centre of each unit cell face.

Regularly-shaped crystals are rare. Chief forms: {100}, {111}, {110}. Penetration twins occur on (111), sometimes in the form of crystalline dendrites. Aggregates. Irregular platy dendrites are observed often. Less frequently there are found whole plates formed in the course of exogenous processes in fissures in rocks. Solid masses weighing as much as several tons have been discovered.

Native copper dendrite

Hardness, 2.5-3. Malleable. Hackly fracture. Cleavage, none. Specific gravity, 8.5-8.9. Other properties. Excellent conductor of electricity. Conductivity 99.95 (100 for silver).

Diagnostic Features: Easily recognised by its colour, malleability, and high specific gravity. Fusible in the blowpipe flame (melting point 1080°C). Readily soluble in dilute HNO_3. Soluble with difficulty in HCl, yielding copper chloride.

Aqueous solution in ammonia is characteristically blue in colour.

Genesis and Occurrence: Native copper is formed in reducing conditions in the course of different geological processes.

Typical hydrothermal deposits of economic importance are very rare. An example are the large native copper deposits in the region of Lake Superior (Michigan, USA), where copper impregnations are associated with zeolites, calcite and other minerals. Native copper is often observed as microinclusions in basic igneous rocks altered by hydrothermal processes, in which it may be the product of decomposition of copper sulphides. Native copper most often occurs in the lower parts of oxidised zones of copper sulphide deposits, in association with cuprite (Cu_2O), malachite, occasionally chalcocite (Cu_2S), and other copper minerals.

It is found as plates or irregular platy dendrites in fissures in host rock in the immediate vicinity of the deposits. Huge tabular slabs of native copper are exhibited at the Leningrad Mining Museum and the Moscow Geological Institute. These came from some old mines in the Degelen Hills (Kazakhstan) where masses weighing several tons had been found. Copper also occurs in sedimentary rocks, mostly sandstones, as binding material between the sand particles, or as irregularly-shaped concretions, occasionally associated with cuprite, malachite, azurite, etc., without an apparent association with primary copper deposits. The genesis of such deposits still remains obscure. Of definite interest are the accumulations of irregularly-shaped copper concretions in the sandstones of the Naukat deposits in Central Asia.

Remarkable twins and dendrites from the upper levels of the well-known Turya copper mines (Northern Urals) were described in 1837 by G. Roze.

Finally, there is the so-called cement copper which is found precipitated on iron objects (cramps, bolts, rails, etc.) in abandoned mine workings flooded with water carrying copper sulphates in solution. The ability of metallic iron to precipitate copper from solution is used for obtaining the latter commercially. Iron filings are placed in special concrete reservoirs which are periodically filled with cupriferous mine waters. When exposed to oxygen, native copper oxidises on the surface. Exposure to water and air results in the formation of hydrocarbonates (malachite, azurite).

Pseudomorphs of native copper occur after cuprite, chalcocite, and occasionally after organic remains (usually fragments of wood).

Uses: The numerous uses of copper as a metal are well known (electrical engineering, machine building, various instruments, utensils, etc.). The process of extracting native copper from its ores is quite simple (by gravity concentration). Therefore, such ores are regarded as economic at much lower tenor than in the case of sulphide and oxidised ores (down to 0.5 per cent for large deposits).

Silver, Ag. Native silver is not as common in nature as copper, and incomparably rarer than gold.

Apart from chemically pure silver, there are varieties: *ktistelite* containing up to 10 per cent and more of isomorphously admixed gold, *copper* silver, *dyscrasite,* etc.

Regularly shaped crystals are rare. Forms {100}, {111}. Twins on (111). Aggregates. Sometimes as typical pinnate dendrites, thin irregular plates and scales.

Also typical are mossy, hair-like, and wire-like forms. Most common are irregularly-shaped grains, and larger nuggets.

Silver-white on freshly broken surface. Often tarnished black. Streak, metallic. Lustre, typically metallic.

Hardness, 2.5. Highly malleable. May be beaten into very thin sheets. Fracture, hackly. Cleavage, none. Specific gravity, 10.1-11.1. The best conductor of heat and electricity.

Diagnostic features: Recognisable by colour, typically hackly fracture, malleability (easily scratched with the point of a knife), and by specific gravity. Distinguished from platinum by lower hardness and specific gravity. Argentite, Ag_2S, a common accessory of native silver, is darker, lead-grey or black in colour.

Fusible in the blowpipe flame. Soluble in nitric acid giving on addition of hydrochloric acid a curdy precipitate of AgCl Blackens on contact with H_2S.

Genesis and Occurrence: Native silver forms in nature very much like copper. Together with other argentiferous minerals, it occurs in *hydrothermal* vein deposits in association with argentite (Ag_2S) and calcite (Kongsberg, Norway), sometimes with complex sulphides, arsenides and antimonides of various metals, including nickel and cobalt.

Under exogenous conditions, native silver, just as native copper, occur in the *oxidised zone* of sulphide and arsenic-antimony ore deposits, being the product of decomposition of the latter or of reduction from surface solutions by various organic compounds. Native silver thus formed is often in the shape of dendrites, plates, and mossy, and filiform aggregates. It was shown experimentally that fine filiform and dendritic formations, which sometimes appear in fine patterns, are precipitated from solution on lumps of coal, especially in the presence of soluble organic compounds.

Under surface conditions, native silver is less stable than gold and is often coated with black gouges or films. In localities with a hot arid climate it frequently alters to stable halides (AgCl, etc.).

Uses: The chief uses of silver are in alloys with copper to make silverware, coinage, and for other purposes. Pure silver is used for filigree work, in the manufacture of crucibles for fusing alkalis, in silver plating, for obtaining chemical compounds, and many other purposes. Eighty per cent of the world's silver production comes as a by-product from silver-rich lead-zinc, gold, and copper ores.

Gold, Au: The most widely distributed member of the group

of native elements; gold has been known to man from time immemorial.

Chemical Composition: Chemically pure gold occurs very rarely. The so-called native gold nearly always contains an isomorphous admixture of silver (usually 4 to 15 per cent by weight). Some varieties containing more silver are classed as a separate mineral species.

The varieties include *copper* gold (cuproaurite) with a Cu content of up to 20 per cent (by weight); *porpezite,* palladium gold, with a Pd content of 5 to 11 per cent and up to 4 per cent of Ag; *bismuth* gold (bismuthaurite) containing up to 4 per cent bismuth in solid solution.

Rare crystals are commonly octahedral {111} 68), less often rhombododecahedral {110}, and occasionally cubic {100}. Faces generally dull, and uneven, sometimes with combination striations parallel to (111) : (310) and (111) : (110) edges. Intergrowths and twins are common on (111). Usually as irregularly-shaped grains in quartz or ore masses. The grains vary greatly in size, but are most often microscopic and sometimes hardly visible, even in polished sections. In valley placers, rounded nuggets are often found weighing several grams, and occasionally tens of kilograms. In the weathering zone of primary deposits smaller dendritic gold nuggets of secondary origin are found occasionally. In the ores of primary deposits dendritic crystalline intergrowths and of reticulate patterns are found occasionally along with small crystals.

Colour, golden-yellow (pale-yellow in the case of argentiferous varieties). *Streak,* metallic, yellow. *Lustre,* typically metallic. *Hardness,* 2.5 to 3.0. Malleable and ductile. May be easily beaten into very thin sheets. *Cleavage,* none. *Specific gravity,* 15.6 to 18.3 (19.30 in the case of pure gold). Good conductor of heat and electricity.

Diagnostic Features: Characteristic golden-yellow colour, low hardness (can be easily cut with a knife), high malleability and specific gravity, does not oxidise on exposure to air.

Distinguished from similar pyrite, chalcopyrite ($CuFeS_2$), and millerite (NiS) by strong lustre and characteristic tint.

Fusible in the blowpipe flame. Insoluble in acids except aqua regia. Soluble in KCN and reagents evolving free chlorine and bromine. Hot high-sulphur alkalis dissolve gold with the formation of sulphides.

Genesis and Occurrence: The bulk of gold occurs in typical *hydro-thermal* deposits genetically associated with intrusions of acid igneous rocks. Paragenetically, it is most often associated with quartz and sulphides (pyrite, arsenopyrite, tetrahedrite, chalcopyrite, less often with galena and sphalerite) and occasionally with tellurides of gold and silver, etc. So-called "free milling gold" is, as a rule, one of the last minerals formed, often being confined to cracks in the earlier minerals. Besides free milling gold, there is also "bound" gold the presence of which in appreciable quantities in sulphides, mainly pyrite and arsenopyrite (FeAsS), is determined by chemical analyses, only some of it being visible under the microscope. Apparently, a part of it is disseminated in a finely dispersed state, similar to the dispersed phase in crystallosols.

Native gold of recent formation is present in the oxidised zone of sulphide deposits, in association with limonite, azurite, the ochres of lead, bismuth, antimony, etc. The silver present in gold is partly removed by surface weathering, which explains the higher purity of gold on the periphery of the deposits and in fissures. It is an old observation that placer gold contains less silver than gold from primary deposits.

The USSR has numerous placer and primary deposits. The biggest nugget in Russian history found in 1842 weighed 35 kg. Of the deposits in other countries, widely known are the gold fields of Transvaal, South Africa, which accounted for nearly 40 per cent of the world's production of gold. Gold occurs there in metamorphosed ancient conglomerates that apparently had formerly been placers. Some of the mine workings there are more than 2 km deep.

Much gold is also produced in the USA (the Western States

and Alaska) and in Canada. The deposits are mostly confined to quartz veins in which gold is generally associated paragenetically with sulphides. Vast gold placer provinces are known in California and Alaska (along the Klondike river, a tributary of the Yukon, and the Seward Peninsula), and in Australia. It was the placers of Victoria (Australia) that yielded the world's biggest nuggets: "Pleasant Stranger" weighing 59.67 kg (1858) and "Desirable" weighing 68.08 kg (1869).

Uses: Gold is the basic metal of currency and coinage. It is also used in jewellery, physical and chemical instruments, dentistry, etc. The minimum gold content of ore that makes mining a commercial proposition ranges from 1 to 10 g per ton, i.e., from 0.0001 to 0.001 per cent (depending on the size of the deposit and the general economic conditions of its development).

This mineral species embraces varieties of intermediate composition in the isomorphous (Au-Ag) series. Silver content exceeds 15 per cent, usually about 30 per cent, sometimes 40 per cent, and even as much as 50 per cent. Cu, Fe, etc., are present in minor quantities. The physicochemical properties are likewise intermediate. Colour, light-yellow to silver-white. Lustre, metallic. Reflectivity very high. Hardness, 2 to 3. Malleable and ductile. Cleavage, none. Specific gravity, 12 to 15.

Occurs almost exclusively in hydrothermal deposits, but much more rarely than native gold and silver. Electrum is typically found in paragenetic association with argentiferous sulphides (argentite, tetrahedrite, proustite, pyrargyrite, etc.). Owing to its high silver content, electrum is lighter than gold and is subject to alteration by weathering: sometimes it is coated with films of halogen and sulphide compounds from which a thin film of native silver is formed under reducing conditions.

IKON-PLATINUM GROUP

The group comprises native metals found in the periodic table under Group VIII: Fe, Co, Ni, Rh, Pd, Ir and Pt.

By the criteria of occurrence and crystal chemistry, the group falls into two subgroups:

(a) The iron sub-group, and

(b) The platinum sub-group.

The minerals of the *native iron* sub-group, which occur in the earth's crust are by the token of their origin classified as those of cosmic origin, i.e., iron meteorites, and exceedingly rare telluric (terrestrial) formations.

The minerals of the *native platinum* sub-group comprising a fairly large number of mineral species and varieties are solid solutions of such metals as Pt, Fe, Ir, Pd and Rh, sometimes of Ni and Cu, rarely of Au, Os and Sn (though sometimes in large amounts) and also of Pb, Zn, Ag, Co, Mn, Mo (up to 0.006 per cent) and Re (0.00008 to 0.002 per cent).

Polyxene and *palladium platinum* are the most common of this latter sub-group. That is loosely called "native platinum" in most cases is polyxene.

A characteristic feature of the minerals of this sub-group is that they do not include isomorphous admixtures of Ru and Os. Although all the six elements of this group have many common physical and chemical features, they behave somewhat differently under natural conditions. In each triad (Pd and Pt) the elements with the maximum atomic weight on the extreme right of the chart behave very differently from Ru and Os on the extreme left. It is still not known whether these two extreme pairs of elements can isomorphously replace one another, though the minerals in which they are found may be in close paragenetic relationship and not infrequently form intergrowths.

The elements Rh and Ir which occupy an intermediate position play a dual role: on the one hand, they form chemical compounds of variable composition with Ru and Os, and on the other hand, they often enter into solid solutions with Pd and Pt and less often with Fe, Ni and Cu. It should be noted that the right-hand and the left-hand pairs of the platinum elements are capable in rare cases of forming natural compounds with S, As, Sb, and partly with Te, whereas no such compounds have as yet been found for Rh and Ir. It is no less significant that while the minerals of the Pd-Pt series, as a rule, crystallise in the cubic system (copper-type crystal structure), the minerals of the Ru-Os series clearly appear as compounds of variable composition crystallising in the hexagonal system (in osmium-type crystal structures).

It is because of that the minerals formed by the platinum elements are divided into two parts, one classed with the iron group, and the other (osmium-ruthenium) as an independent group.

IRON

Chemical composition. According to chemical analysis, telluric iron is almost pure iron with minor admixtures of: Ni up to 0.6 per cent, sometimes up to 2 per cent, seldom more; Co up to 0.3 per cent; Cu up to 0.4 per cent; Pt up to 0.1 per cent.

Occurs extremely rarely in very small crystals. Usually as minute irregularly-shaped grains, less often in larger accumulations. Several polymorphous modifications of iron have been established. Of these y-Fe, a high-temperature modification (above 906°C) has the structure of a face-centred cube of the copper-type (a_0 = 3.63), and a-Fe, a low-temperature modification, has the structure of a centred cube of the a-Fe type (a_0 = 2.86).

Colour, steel-grey, in polished sections metallic white. *Streak,* shiny, steel-grey. *Lustre,* typically metallic on fresh fracture. *Hardness,* 4 to 5. Malleable. *Cleavage,* on {100}. *Specific gravity,* 7 to 7.8. Strongly magnetic.

Diagnostic Features: Distinguished from native platinum by solubility in nitric acid, lower specific gravity, strong magnetism and ready oxidation in air. Nickel-poor varieties precipitate metallic copper onto their surface from a solution of copper sulphate. When polished and etched, the surface of meteoric iron usually shows a coarse-latticed structure, never observed in telluric iron.

Genesis and Occurrence. The rare occurrences of telluric native iron are confined to basic and ultrabasic igneous rocks. Larger masses have been found in the basalts at Uifak on Disko Island, Greenland, and near Kassel, Germany. In both localities it is associated with pyrrhotite (FeS) and cohenite (iron carbide, Fe_3C). It has been repeatedly found as microscopic grains in altered (serpentinised) ultrabasic rocks, also product of the reduction reactions of the latter.

There are also indications of its formation by exogenous processes as impregnations in siliceous schists on the island of Kalimantan (Borneo), in the products of coal fires and in peats.

Widmanstatten Figures on a Polished Surface

Polyxene—(Pt, Fe)—The name originates from its numerous isomorphous admixtures; "poly" is the Greek for many and "xe-nos", foreign. Polyxene is the most widespread platinum mineral in the earth's crust.

Chemical Composition: Pt 80 to 88 per cent, Fe 9 to 11 per cent, sometimes diminishing to 4 or 5 per cent or increasing over 11 per cent (with corresponding changes in the Pt content). Isomorphous admixtures: Ir up to *1 per cent — iridioplatinum* Pd 0.1 to 1.0 per cent, sometimes up to, 7 per cent and more—*palladium platinum;* Rh 0.1 to 0.5 per cent, sometimes as high as 4 to 5 per cent—*rhodium platinum]* Cu up to 0.8 per cent; Ni—from traces up to tenths of a per cent, sometimes in sizable quantities—*nickel platinum.*

Usually irregularly shaped grains; rare small crystals are predominantly cubic in form 72). The combination of faces, besides the dominant form {100}, include {110}, {210}, {310} and some others. Penetration twins with (100) as the twinning plane and contact twins on (111) are the most frequent forms of twinning. Individual grains of native platinum occurring in ores often form small clusters, and sometimes solid masses or nuggets.

Structure: Several polymorphous modifications of iron have been established. Of these y-Fe, a high-temperature modification (above 906°C) has the structure of a face-centred cube of the copper-type (a_0 = 3.63), and a-Fe, a low-temperature modification, has the structure of a centred cube of the a-Fe type (a_0 = 2.86).

Colour, steel-grey, in polished sections metallic white. *Streak*, shiny, steel-grey. *Lustre*, typically metallic on fresh fracture. *Hardness*, 4 to 5. Malleable. *Cleavage*, on {100}. *Specific gravity*, 7 to 7.8. Strongly magnetic.

Diagnostic Features: Distinguished from native platinum by solubility in nitric acid, lower specific gravity, strong magnetism and ready oxidation in air. Nickel-poor varieties precipitate metallic copper onto their surface from a solution of copper sulphate. When polished and etched, the surface of meteoric iron usually shows a coarse-latticed structure, never observed in telluric iron.

Genesis and Occurrence: The rare occurrences of telluric native iron are confined to basic and ultrabasic igneous rocks. Larger masses have been found in the basalts at Uifak on Disko Island, Greenland, and near Kassel, Germany. In both localities it is associated with pyrrhotite 427.5 g. Nuggets found in placers were sometimes as big as 10 x 18 cm and weighed from 8 to 9 kg.

Colour, silver-white to steel-black. Streak, metallic, steel-grey. Lustre, typically metallic. *Hardness*, 4 to 4.5 and as high as 6 to 7 in iridium-rich varieties. Malleable. Hackly fracture. *Cleavage*, usually none. *Specific gravity*, 15 to 19. It has been observed that the lower specific gravity is due either to the presence of cavities filled with natural gases or to inclusions of foreign minerals. Other properties. Magnetic. Good conductor of electricity.

Diagnostic Features: Outwardly polyxene most closely resembles native silver and native iron. It is distinguished from native silver by greater hardness and higher specific gravity, by being infusible in the blowpipe flame and insoluble in acids (except aqua regia). The latter property also distinguishes it from native iron.

Genesis and occurrence: The platinum minerals, in the majority of eases, occur in typically magmatic deposits genetically related to *ultra-basic* igneous rocks. They are among the last to crystallise in ore bodies (after silicates and oxides) at the hydrothermal stage of the magmatic process.

Palladium-poor platinum minerals (polyxene, iridioplati num, etc.) occur in deposits in dunites—olivine non-feldspatic rocks high in magnesia and low in silica. Paragenetically, they are very closely associated with chromè-spinellids which are oxides of complex composition: $(Fe, Mg)(Cr, Al, Fe)_2O_4$–.

Palladium platinum and nickel-palladium platinum are, for the most part, found in *basic* igneous rocks (norites, gabbro-diabases), generally in association with sulphides: pyrrhotite (FeS), chalcopyrite ($CuFeS_2$), and pentlandite $(Fe, Ni)_9S_8$.

Under exogenous conditions the disintegration of rocks and primary deposits gives rise to platinum placers. The majority of the minerals of the sub-group are chemically inert under these conditions. Along with gold, platinum has been known to man since ancient time. In Europe, platinum attracted attention in 1735 when Spanish mathematician Antonio de Ulloa, who had travelled in South America, brought from Colombia metallic grains resembling silver in colour, but strongly differing from the latter in other properties.

In the Urals, native platinum first attracted attention as an admixture to placer gold in 1819. Later, rich platinum placers of world importance were found in the *Middle* and *Northern Urals,* where they were confined to outcrops of ultrabasic rock masses (dunites and pyroxenites). Numerous small primary deposits occur in the dunite mass of *Nizhny Tagil* (the Urals). Native platinum (polyxene) accumulations, more often than not, are confined to chromite ore bodies composed of chrome-spinellids with an admixture of silicates (olivine and serpentine).

At the present time, a major supplier of platinum metals for industry and commerce is the ore deposit at Sudbury (Canada) from which they are extracted as a by-product of nickel, copper and cobalt.

Uses: Originally, not only was platinum rejected as useless, but was considered to be a harmful impurity of placer gold with which it was extracted. It was either discarded with the tailings of gold washing or used by hunters for shot. Sometimes, attempts were made to gild and pass it for gold. Thus, the first articles made of native platinum in the Urals, later displayed at the Leningrad Mining Museum, were crude chains and rings, barrel hoops, etc. It was only later that the outstanding properties of the platinum metals were discovered.

These valuable properties include refractoriness, electrical conductivity and chemical inertness, all of which make platinum an important material in the manufacture of laboratory equipment, the production of sulphuric acid, in electrical engineering, etc. A great deal of platinum is consumed in jewellery and dentistry.

"Crude" platinum is delivered to special refineries where it is separated into its pure constituent metals.

Osmium-Ruthenium (Osmibidium) Group

These two elements are found on the extreme left of the triads of the platinum group elements in the periodic classification. In nature they characteristically isomorphously replace one another and readily form solid solutions with the elements in the middle of these triads, i.e., with rhodium and especially with iridium, but not with palladium or platinum. The presence of platinum as an element, though often indicated by chemical analysis of the minerals of the group, is probably limited to mechanical impurities, which is in some cases confirmed by the microscopic examination of polished sections. Thus, the principal metals of the group are osmium and iridium and, to a lesser degree, ruthenium and rhodium.

Although found together with the minerals of the iron-platinum group, the minerals of this group sharply differ from them in many properties such as: (1) characteristically tabular crystal habit; (2) pronounced optical anisotropy; (3) definitely higher hardness due to which the osmiridium platelets stand out in polished sections against the background of native platinum; (4) a relatively high specific gravity; and (5) an unusual chemical inertness (insoluble even in boiling aqua regia).

The principal species are generally distinguished by their osmium-to-iridium ratio and content.

Nevyanskite: *(Ir, Os):* Named after its place of discovery in the Nevyansk district north of Sverdlovsk (the Urals). The most widespread mineral variety of this group in nature.

Chemical Composition: The minerals described under the general name of nevyanskite vary in the content of individual metals as follows (per cent): Ir 46.8 to 77.2, Os 21.0 to 49.3, Ru 0 to 0.5, Rh 0.5 to 7.7, Pt 0.1 to 5.5, Cu 0 to 0.9, Fe 0 to 1.4.

The Ir: Os ratio varies from 4:1 to 1: 1. Sometimes there is a large proportion of rhodium isomorphously admixed with iridium. System hexagonal. Crystal structure hexagonal, closest packing. In contrast to the minerals of the iron-platinum group, the varieties of this group are very well individualised crystallographically. Due to the predominant {0001} pinacoidal development, the crystals are usually small hexagonal plates or tabular grains bounded at the sides by {1010} prism and {1011} hexagonally-dipyramidal faces. Hexagons with unequal sides, approximately triangular in shape, have been observed. Intergrowths of platy individuals fairly frequent.

Colour, tin-white to light-grey. Lustre, metallic. Optically anisotropic. Hardness, 6 to 7. Brittle. Cleavage, {0001} perfect, but the folia are-not easily separated. Folia friable. Specific gravity, 17.0 to 21.0. Low specific gravity is caused by the presence of small gas-filled vesicles, sometimes very numerous (up to 17 per cent by volume).

Diagnostic Features: Easily recognisable by platy habit, lighter colour (as compared with syssertskite) and higher hardness. Infusible in the blowpipe flame. When fused with nitre, gives off fumes of osmic anhydrite (OsO_4) with a pungent odour. When the fused mass is dissolved in water, a black iridium powder is precipitated. Highly resistant to acids.

Genesis and Occurrence: Minerals of this group are for the most part genetically associated with igneous ultrabasic rock masses (dunites and peridotites) in which they are found in close association with the platinum-group minerals, chrome-spinellida, occasionally with copper sulphides. They may also occur in the rocks, i.e., are not accompanied by other ore materials.

Occurrences have been recorded from hydrothermal auriferous quartz veins. Nevyanskite intergrowths with gold have also been found. Due to its chemical inertness nevyanskite, together with gold and native platinum minerals, passes into placers deriving from the weathering and erosion of primary deposits.

Osmiridium occurs in placers confined to the outcrops of ultrabasic igneous rock masses. It is found as insignificant admixture in many gold placers in the Urals, Siberia, and also in California and Oregon (USA). On the island of *Tasmania* placers are known in which the osmiridium minerals prevail over the platinum metals. Osmiridium is also extracted as a by-product at the *Witwatersrand* gold fields (Transvaal, South Africa).

Uses: The larger crystals (at least 1 mm in diameter) of osmiridium are used in the natural state (without chemical treatment) in the manufacture of instruments for special purposes (grain no less than 0.5 mra in size, preferably oval, not platy, are required for this purpose). Smaller grains are used for fountain-pen nibs, the cutting parts of surgical instruments, etc.

Syssertskite, (Os, Ir). Named after the place of discovery—

the Syssert district (near Sverdlovsk) in the Urals. Chemical composition. In syssertskite osmium prevails over iridium. Ru is often present in considerable quantities, replacing osmium.

System, hexagonal; symmetry, dihexagonal dipyramidal. Syssert-skitehas been observed in hexagonal platy crystals, rounded plates and fragments (in platiniferous placers). No special study of its crystals has been made.

Colour, steel-grey to dark-grey. Lustre, dull metallic.

Hardness, about 6. Brittle. Cleavage, {001} distinct. Specific gravity varies from 17.8 to 22.5. Magnetic properties unstudied. Diagnostic features distinguished from nevyanskite only by darker colour. Other distinguishing features can be established only by chemical analysis.

Infusible in the blowpipe flame. On thorough ignition in an oxidising flame, becomes black, giving off vapours of osmic anhydrite (OsO_4) of pungent unpleasant odour. Fused with nitre, forms a green mass which on boiling gives a black precipitate of iridium. Highly resistant to acids.

Group of Semi-metals

Besides arsenic, this group includes antimony and bismuth, i.e., the elements of the long periods of Group V of the periodic table. All of them are found native, though rarely. They crystallise in the same system and form crystal structures of the same type. However, they do not occur together nor do they form solid solutions or compounds of definite composition, with the exception of arsenic and antimony which, at high temperatures form solid solutions in all proportions, while at low temperatures only AsSb (allemontite), a stable intermetallic compound, is formed.

The structural peculiarity of the semi-metals under consideration is that each atom is bonded by covalent bonds to three out of the six atoms surrounding it, thus forming sheets of closely packed structural units. Between the layers, the bonding is by van der Waals forces, i.e., is very weak, which results in perfect cleavage along (0001), high optical anisotropy and low hardness.

Arsenic, As. Found comparatively rarely and usually in small quantities.

Chemical composition: Arsenic content 84 to 98 per cent. Impurities: Sb 1.7 to 9.2 per cent, less often Ag, Fe and Ni, occasionally Bi and V. It is possible that the content of Ag, Ni, Fe and Bi is due to mechanical inclusions of foreign minerals (native bismuth, and nickel and iron arsenides with which native arsenic is sometimes closely intergrown).

System, trigonal; symmetry, ditrigonal scalenohedral Extremely rare crystals, rhombohedral or pseudocubic. Aggregates. Usually occurs in crusts with a reniform surface; stalactitic, conchoidal formations which show a crystalline granular structure on fracture.

Colour, tin-white on freshly broken surface. On exposure to air it soon turns yellow-brown and then black. Streek, grey, tin-white. Lustre, metallic, strong on fresh fracture, soon grows dull, and with time blackens due to surface oxidation.

Hardness 3.5. Brittle. Cleavage {0001} perfect and (0112) less perfect. Fracture granular. Specific gravity 5.63 to 5.78.

Diagnostic Features: Native arsenic is rather easily identified by its form, black-tarnished surface, high specific gravity, strong metallic lustre on fresh fracture and perfect cleavage.

Volatilises in the blowpipe flame (at about 360°C) without fusion, giving off a peculiar garlic odour, and leaves a white coating on charcoal. Liquefies only under high external pressure. In the closed tube gives an arsenic mirror. Smells of garlic when struck with a hammer.

Genesis and Occurrence: Arsenic is found in *hydrothermal* deposits in cavities as metacolloidal formations, and is evidently formed at the very last stages of hydrothermal activity. May occur in association with diverse arsenides, antimonides, less often sulphides of nickel, cobalt, silver, lead, etc., as well as non-metallic minerals.

Secondary formation of arsenic in the weathering zones of arseniferous ore deposits mentioned in literature seems highly unlikely because under the circumstances, it is very unstable and rapidly oxidises with complete decomposition. The black crusts formed constitute a mixture of arsenic and arsenolite (As_2O_3), which, in the end, turn into pure arsenolite.

In the USSR, native arsenic has been found in a few deposits. As reniform masses on crystalline calcite together with galena and sphalerite it has been repeatedly recorded from the *Sadon* hydrothermal lead-zinc deposit. Large reniform masses of native arsenic of a con-centric-conchoidal structure were found on the left bank of the *Chikoy* river (Trans-Baikal Region). The only mineral observed in paragenesis with arsenic was calcite occurring as fluccans on the walls of thin veins cutting across ancient crystalline schists. Arsenic fragments have also been found in the area of *Dzhalinda* Station of the Amur railway, and elsewhere.

In a number of deposits of Saxony (Freiberg, Schneeberg, Annaberg, etc.) native arsenic was found in association with arsenious compounds of cobalt, nickel, silver, native bismuth, etc. These and other finds of native arsenic are of no practical value.

Bismuth, Bi: Although native bismuth occurs much more frequently than arsenic and antimony, it is a rare mineral nevertheless and usually is not found in large accumulations.

Chemical Composition: Generally bismuth is almost pure. However, there may be insignificant impurities or rather traces of Fe, As, S, and Sb.

System trigonal; symmetry ditrigonal-scalenohedral Extremely rare crystals are poorly developed. More frequently occurs as impregnated grains, sometimes as large, lamellar or granular aggregates, and also as plumose dendrites.

Colour on fresh fracture, silver-white with a yellow tinge;

with time a peculiar reddish tarnish appears. Streak, grey. Lustre, metallic, strong. Hardness, 2.5. Slightly malleable. Cleavage, {0001} highly perfect and {1011} distinct. Specific gravity, 9.70 to 9.83 (when liquid 10.03). Melting point about 270°. Boiling point about 1450°. Diamagnetic.

Diagnostic Features: Easily recognisable by slight yellow-reddish tarnish, strong metallic lustre, perfect cleavage, relatively low hardness and relatively high specific gravity.

Readily fusible in the blowpipe flame. On prolonged blowpiping vaporises giving white coatings on charcoal which change to orange-yellow and become lemon-yellow on cooling. Fused with KI and S, gives a characteristic coating on charcoal with a bright-red fringe of BiI_3. Readily soluble in nitric acid, giving a white precipitate upon dilution.

Genesis and Occurrence: Bismuth is almost invariably found in formations of hydrothermal origin.

In some instances it is genetically associated with high-temperature formations, occasionally with pegmatites but more often with tin-tungsten deposits—in association with such minerals as arsenopyrite (FeAsS), bismuthite (Bi_2S_3), wolframite ($FeWO_4$), molybdenite (MoS_2), cassiterite (SnO_2), topaz, beryl, tourmaline, etc.

In other instances, native bismuth is closely associated with the arsenious compounds of nickel and cobalt, uranium minerals, sulphur compounds of silver, lead and zinc, etc. Examples are the associations in the numerous veins of the Erzgebirge mountains (at the border between Saxony and Bohemia) in the vicinity of *Schneeberg, Annaberg* and *Jachymov*. Native bismuth occurs in larger quantities together with bismuthite (Bi_2S_3) in the quartz-barite veins of the *Tasna* and *Oruro* in Bolivia, and elsewhere.

Sulphur Group

This group comprises native elements of Group VI of the periodic table: sulphur, selenium, and tellurium. Of these sulphur is a typical metalloid, whereas selenium and especially tellurium possess certain properties of semi-metals.

Also three natural polymorphous modifications of sulphur are known, of which only the orthorhombic a-sulphur is stable in natural conditions. At atmospheric pressure and temperatures over 96.5°, it changes to J-sulphur in the monoclinic system and on cooling resumes its former crystallinity; y-sulphur, although monoclinic, is not stable at atmospheric pressure and low temperatures, and at room temperature it turns into a-sulphur. Three other polymorphous modifications of sulphur obtained artificially never occur in nature.

Sulphur, S. a-sulphur, the most stable form at room temperature, is generally referred to as orthorhombic sulphur of simply sulphur.

Chemical composition: Occasionally native sulphur is chemically pure, but usually it is contaminated with mechanical impurities such as clayey or organic matter, oil drops, gases, etc. Also known are rare varieties with an isomorphous admixture of Se, usually up to 1 per cent, less often up to 5.2 per cent in the form of *selenosulphur,* and also with Te, As, and, in exceptional cases, with Ti.

System, orthorhombic; symmetry, rhombic-dipyramidal. Crystal structure. Orthorhombic sulphur, according to X-ray data, has a *molecular* and very complex lattice which is uncommon in inorganic substances. In the crystal structure each atom on two sides has spheres which intersect with the spheres of the adjacent atoms, chains composed of 8 atoms forming closed zigzag-shaped or "puckered" rings. Hence it is evident that the sulphur molecule is S_8. The distance between the S —S atoms is 2.12 Å. The unit cell is composed of 16 such electrically neutral molecules (rings) very loosely held together by van der Waals forces. Commonly pyramidal or truncated pyramidal, less commonly rhombotetrahedral. Characteristic forms: {001}, {Oil}, {111}, {113}, etc. Twins are rare with (111), sometimes (011) and (110) as the composition plane. Aggregates. Often massive or earthy. Occasionally in sinter reniform forms and coatings (in areas of active volcanoes).

Colour, a-sulphur may have various shades of yellow: straw yellow, honey yellow, yellowish grey, brownish, and black (due to carbonaceous impurities). Streak, almost white, though powder is pale yellow. Lustre, adamantine on faces, greasy on fracture. Crystals translucent.

Hardness, 1 to 2. Brittle. Cleavage {001}, {110} and {111} imperfect. Specific gravity, 2.05 to 2.08. Very poor conductor of electricity and heat (good insulator). Becomes negatively electrified on rubbing. May decrepitate from the warmth of the hand.

Diagnostic Features: Recognisable by colour, low hardness, brittleness, greasy lustre on fracture and low melting point. Easily melts before the blowpipe and even in the flame of a match (at 112.8°); burns with a blue flame and with the characteristic SO_2 odour (which distinguishes it from similar auripigment, As_2S_3). Readily soluble in carbon disulphide, turpentine and kerosene, but does not decompose in hydrochloric and sulphuric acids. Strong nitric acid and aqua regia oxidise sulphur to H_2SO_4.

Being the only mineral in this section which possesses a *molecular* structure, native sulphur is altogether unique in its properties. The presence of electrically neutral S_8 molecules as structural units explains such properties as poor electrical and heat conductivity, loose intermolecular bonds with the consequent low melting and sublimation points, low mechanical strength, low hardness and indistinct cleavage, due to which it gives very uneven fracture and shows a greasy lustre. The circular shape of the structural units (S_8 molecules) which is strikingly different from the spherical shape also determines the pronounced optical anisotropy of the crystalline matter, the anisotropy of heat expansion, etc.

Such properties as low specific gravity and melting point

are taken advantage of in the recovery of native sulphur from deep-seated sulphur bearing strata. Superheated steam is injected into boreholes, through pipes, and the molten sulphur which incidently drops its impurities rises with the upward stream to the surface where it is collected by most simple methods.

Genesis and Occurrence. Native sulphur occurs exclusively in the uppermost parts of the earth's crust or right on its surface. It is formed in various ways.

During volcanic eruptions, as sublimates on the walls of craters and in rock fissures. Sometimes molten sulphur is carried out with hot waters of solfataric springs and solidifies upon cooling in gullies and valleys in regions of modern active volcanoes (the Kurile Islands, Japan). It is often associated with other sublimates as well as with sulphates formed by the action of SO_2 on enclosing rocks. It is also formed through incomplete oxidation of hydrogen sulphide in solfataras or as a product of the reaction of H_2S with sulphur dioxide:

$$2H_2S + 2O \rightarrow 2H_2O + 2S; \; H_2S + SO_2 \rightarrow H_2O + O + 2S$$

Through the decomposition of metal sulphides, chiefly pyrite, in the lower part of the zone of oxidation of ore deposits. It is usually strongly contaminated with various mechanical impurities.

Through the decomposition of gypsiferous sedimentary strata. Often occurs in paragenesis with gypsum on the decomposed parts of which it forms crystalline or powdery masses. The mechanism of this process is still not very clear.

By sedimentation (biochemically) in normal sedimentary rocks usually in beds containing gypsum, solid and liquid bitumens (asphalt, petroleum), etc. Deposits of this type are widespread all over the globe and are of great commercial importance. The bulk of native sulphur apparently has formed syngenetically, i.e., simultaneously with other sediments. However, not infrequently there are found epigenetic sulphur formations in host rocks, due apparently to later migration caused by tectonic processes, transportation by oils, solutions, etc.

The biochemical genesis of sulphur is also linked with the life activity of anaerobic bacteria, which results in the evolution of hydrogen sulphide, the incomplete oxidation of which results in precipitation of sulphur.

In the Soviet Union, numerous sulphur deposits are found in Central Asia.

Outside the USSR, best known are the huge deposits in *Sicily* and those associated with salt domes in *Texas* and *Louisiana* (USA).

Uses: The most important use of sulphur is in the manufacture of sulphuric acid which is vital to many industries; it is also used in agriculture (pest control), in the rubber industry (vulcanisation), in making matches, fireworks, dyes, etc.

Carbon Group

This group of minerals also holds a special position among the native elements. It comprises two polymorphous modifications of carbon sharply differing in physical properties: diamond and graphite.

The crystal structure of *diamond,* which on the whole is similar to that of a face-centred cube, differs from the latter in that the atoms of carbon are located both on the faces of the cube and also in the centres of half of the small cubes which alternate with small empty cubes.

The general conception is that the carbon atoms in the crystal structure of diamond are linked to each other exclusively by covalent bonds along directions connecting the centre of a tetrahedron with its apexes. However, N.V. Belov has suggested a more natural conception of diamond structure as one of the ZnS type with two kinds of ions, C^{4+} and G^{4-} (the C^{4+} cation radius is about 0.15Å and the C^{4-} anion radius is about 1.5Å), which accounts for the closest packing of the anions in the structure. This concept agrees well with the crystal forms of diamond and its properties such as absence of colour, low electrical conductivity, exceptionally high hardness, high stability within a wide range of temperatures and pressures (in particular it does not show any change when heated up to 2500° C in the absence of oxygen), high resistance to acids and alkalis, etc.

The structure of *graphite* differs strikingly from that of diamond. The ions of carbon in graphite lie in sheets, which are presented by planar hexagonal nets. According to N.V. Belov, the crystal structure of graphite consists of most closely packed large C^{4-} anions with half of the triangles in each sheet occupied by smaller C^{4+} cations. Each ion in the planar net is surrounded by three neighbouring ions at a distance of 1.42Å (1.54Å for diamond), while the distance between the planar nets is more than twice as large (3.40Å). Hence the specific properties of graphite: its much lower specific gravity, as compared with diamond, great ability to split into thin scales, pronounced optical anisotropy, and hardness anisotropy which is revealed only by very accurate and delicate investigation techniques (perpendicular to the cleavage plane hardness is 5.5, but at the same time graphite is so soft owing to the loose cohesion of the sheets so that it soils the fingers and marks paper). The same structural peculiarities explain the non-uniform light absorption of graphite and its consequent black colour.

It is assumed that, in contrast to diamond, the type of bonding in graphite is somewhat metallic, i.e., "metallic" electrons also take part in the bonding. This agrees well with such properties as sub-metallic lustre, good electrical conductivity, etc. In chemical and thermal stability, however, graphite is quite close to diamond.

If we represent the structure of diamond along the trigonal axis (cf. the numbers of the atoms in 81a), we shall also observe hexagonal rings in the horizontal planar nets (atoms 6, 11,8,9,7, and 10) with the difference that the nets are not quite planar:

three of the atoms are somewhat above the other three. That is why diamond crystals have medium cleavage parallel to {HI}, while in graphite crystals cleavage is perfect parallel to {0001}.

Diamond, C: The name is derived from the Greek "adamas" meaning invincible (owing to its supreme hardness and resistance to the action of chemical and physical agents).

Varieties: (1) *bort,* irregular intergrowths and spherical radiate aggregates; (2) *carbonado,* close-grained porous aggregates coloured brownish black by amorphous graphite and other impurities.

Chemical Composition: Colourless diamonds are pure carbon. Coloured and opaque varieties yield on combustion a residue, as high as a few per cent, containing SiO_2, MgO, CaO, FeO, Fe_2O_3, Al_2O_3, TiO_2, etc. Graphite and other minerals are often present as inclusions. System, cubic; symmetry, hexoctahedral.

Habit, octahedral 85, less common dodecahedral; seldom cubic, and occasionally tetrahedral. Characteristic forms: {111}, {100}, and {110}. Curved faces are quite frequent 86) as well as so-called "dodecahedroids" and "octahedroids" described in detail by Soviet scientists I.I. Shafranovsky and A.A. Kukharenko. Penetration twins occur along (111), less often along (100). Crystals range in size from very fine to very large specimens weighing hundreds or even thousands of carats (a metric carat is 0.2 g). The largest stones ever found: "The Cullinan", 3,025 carats; "The Excelsior", 969.5 carats; "The Victoria", 457 carats; "The Orloff", 199.6 carats; etc.

Colour, Colourless water-transparent or light blue, yellow, brown, and black. Lustre strong, adamantine. Refractive index n=2.40 to 2.48. *Hardness,* Absolute hardness is 1,000 times greater than that of quartz and 150 times greater than that of corundum. Brittle. *Cleavage,* (111) distinct. *Specific gravity,* 3.47 to 3.56. Poor conductor of electricity.

Diagnostic Features: Diamond is unique among minerals

in its hardness. Other characteristics are its strong adamantine lustre and frequent curved crystal faces. Fine grains in heavy concentrates are easily recognised by their fluorescence in ultraviolet light. The fluorescence colours are usually blue, sometimes green.

Genesis and Occurrence: Primary deposits are genetically associated with ultrabasic plutonic igneous rocks: peridotites, kimberlites, etc. In these rocks, diamond probably crystallises at great depths at high temperatures and pressures. Judging by its form and mode of occurrence, diamond was among the first (minerals) to crystallise from the magmas. It is not clear whether diamond crystallised from the carbon contained in the magma itself or from that assimilated from the chromespinellids, (Fe, Mg) (Cr, Al, Fe)$_2$O$_4$; ilmenite, ($FeTiO_3$); pyrope (red garnet); magnetite, ($FeFe_2O_4$); and haematite, (Fe_2O_3), etc.

The world's largest diamond deposits are located in a number of areas in South Africa: along the Vaal River, in Southwestern Transvaal, and on the southwestern coast of Africa. In these localities the diamond-bearing rocks are semi-decomposed kimberlites which, together with clastic rocks, fill gigantic vertical volcanic pipes which are several kilometres deep. The pipes which are of elliptical or irregular cross-section have probably resulted from the tremendous explosive processes which once took place at great depths. In some deposits diamonds are recovered from depths exceeding 1 km (Kimberley). The extracted ore ("blue ground") is crushed and washed at concentration plants. The mean diamond content in the ore is 0.000052 per cent by weight. It was in South Africa that the world's largest diamonds were found ("The Cullinan", "The Excelsior", etc.). In addition to mines, there are many rich alluvial placers in South-West Africa, in the basins of the Vaal and Orange rivers. The Congo is another major producer of diamond.

Recently relatively many diamond-bearing pipes have been discovered in Northern Yakutia. These are filled, however, with somewhat more recent kimberlites. Some of the pipes were given such names as "The Peace Pipe", "Summer Lightning", etc. The output of diamonds in Siberia is increasing year by year.

Placer diamond deposits are formed through the weathering and erosion of diamond-bearing rocks. They are found in South Africa, Brazil (Minas Gerais), and India where diamond mining dates to ancient times, and where large diamonds like "The Orloff" and "The Kohinoor" were found. Diamond is also found in some other localities.

Uses: Perfectly transparent diamond is the most valued gemstone. Very small diamonds, and also bort and carbonado are used in metal-working, stonecutting industries, in the production of abrasives, etc.

Graphite, C: The name derives from the Greek "grapho" meaning / write. Varieties: graphitite—cryptocrystalline variety, schungite— amorphous variety, probably derived from the natural coking of coals.

Chemical Composition: Rarely found pure. May contain as much as 10 to 20 per cent of ash comprised of various constituents (SiO_2, Al_2O_3, FeO, MgO, CaO, P_2O_5, CuO, etc.), sometimes also water, bitumens, and gases (up to 2 per cent).

System: hexagonal; symmetry, dihexagonal-dipyramidal. Well-developed crystals extremely rare and are in the form of hexagonal plates or tablets 87), sometimes with trigonal striations on the (0001) face. Characteristic forms: {0001}, sometimes {1011} and {1122}, etc. Aggregates, often scaly. Less often in rod-like or fibrous masses.

Colour, Iron black to steel grey. Streak, shiny black. Lustre, sub-metallic; cryptocrystalline aggregates dull. Optically negative. Refractive index n_y = 1.93-2.07. Transmits grey light in thin folia. *Hardness, 1:* Thin folia flexible. Greasy touch. Marks paper and soils the fingers. *Cleavage* {0001} perfect. *Specific gravity,* 2.09 to 2.23 (depending on the degree of dispersion and the presence of very fine pores), 1.84 to 1.98

in the case of shungite. Good conductor of electricity due to very close packing of atoms in the sheets.

Diagnostic Features: Easily recognisable by colour, low hardness, and greasy touch. Distinguished from molybdenite (MoS_2) by darker iron-black colour, weaker lustre, lower specific gravity (floats in bromoform) and higher electrical conductivity (molybdenite is a poor conductor of electricity). When touched with a zinc bar through a drop of $CuSO_4$, a thin film of copper will be deposited on graphite.

Infusible in the blowpipe flame. When heated in oxygen burns with greater difficulty than diamond. Volatilises without fusing in the electric arc. Insoluble in acids. In powdered form burns with a flash when mixed with KNO_3 and heated.

Genesis and Occurrence: In nature graphite is formed in the course of reduction processes at high temperatures.

Sometimes occurs in *magmatic* rocks of diverse composition. In many instances, the source of carbon are carboniferous enclosing rocks. Graphite is also found occasionally in *pegmatites.* In the provinces of Ontario and Quebec in Canada graphite deposits are found in limestones near the contact with igneous rocks. Vein deposits of coarsely laminated graphite occur in Ceylon.

Very widespread are deposits of graphite formed *metamorphically* from coal or bitumen in the conditions of regional metamorphism or under the action of magmatic intrusions.

In the USSR, vast deposits of graphite are found at *Aliber* in the Tunkin mountains. The graphite is found in the form of stocks in igneous rocks (syenites) and constitutes a mixture of high-grade scaly or finely crystalline graphite with magmatic minerals (microcline, nepheline, etc.). In places these ore bodies are fringed by coarse-grained limestone. The magmatic rocks themselves also contain graphite. Very interesting graphite formations of the secretion type are found within graphite masses. These are irregular small bodies with outer crust of tangentially oriented scales and an internal radial structure. Graphite is believed to be genetically associated with the reduction of the carbon oxide evolved when limestones were engulfed by magma.

Along the northwestern shore of the *Sea of Azov* there are worked deposits of impregnated coarse-scaly graphite occurring in the form of superficially weathered beds in ancient metamorphic rocks (gneisses). An interesting point is that marginal portions of intersecting pegmatite and aplite veins contain coarse-crystalline hexagonal graphite individuals, whereas in the middle of the veins finely-crystalline segregations are present.

Uses: Graphite has a wide variety of applications such as in the manufacture of graphite crucibles, in foundry facings, the manufacture of pencils, electrodes, lubricants, and pigments.

Bibliography

Anthony, John W., Bideaux Richard A., Bladh Kenneth W. and Nichols Monte C.: *Handbook of Mineralogy: Elements, Sulfides, Sulfosalts,* Mineral Data Publishing, 1990.

———— : *Mineralogy: Silica, Silicates,* Mineral Data Publishing, 1995.

———— : *Mineralogy: Halides, Hydroxides, Oxides,* Mineral Data Publishing, 1997.

———— : *Handbook of Mineralogy, Volume IV, Arsenates, Phosphates, Vanadates,* Mineral Data Publishing, 2000.

———— : *Mineralogy of Arizona,* University of Arizona Press, Tucson, 1977.

Artini E.: *Mineral,* Hoepli, 1934.

Aubert, G., Guillemin C. and Pierrot R.: *Précis de Minéralogie,* Masson, 1978.

Bancroft, Peter: *The World's Finest Minerals and Crystals,* Viking Press, New York, 1973.

Bates, Robert Latimer: *Our Modern Stone Age,* W. Kaufmann, Los Altos, 1982.

Bayliss, Peter, Erd Dick C., Mrose Mary E., Sabina Ann P. and Smith Deane K.: *Mineral Powder Diffraction File, Data Book, International Centre for Diffraction Data,* 1986.

Betekhtin, A. G.: *Kurs Mineralogii,* Mir, Moskva, 1968.

Beudant F. S.: *Traité élémentaire de Minéralogie,* Verdière, 1830.

Blackburn, William H. and Dennen William H.: *Encyclopedia of Mineral Names*, Mineralogical Association of Canada, 1997.

Bowen, Oliver E.: *Rocks and Minerals of the San Francisco Bay Region*, University of California Press, Berkeley, 1962.

Chesterman, Charles Wesley: *The Audubon Society Field Guide to North American Rocks and Minerals*, Knopf, New York, 1978.

Clark A. M.: *Hey's Mineral Index*, Natural History Museum Publications, Chapman, 1993.

Dana Edward Salisbury, Ford William E., *A Textbook of Mineralogy*, John Wiley, 1949.

De Michele, Vincenzo: *Minerals, their Beauty and Structure*, Orbis Pub., London, 1976.

Deer, W. A.: *Rock-Forming Minerals*, Wiley, New York, 1962.

Deer, W.A., Howie R.A. and Zussman J.: *An Introduction to the Rock forming Minerals*, Longman, London, 1993.

Desautels, Paul E.: *The Mineral Kingdom*, Madison Square Press, New York, 1968.

——————— : *Treasures in the Smithsonian: The Gem Collection*, Smithsonian Institution Press, Washington, 1979.

Eckel, Edwin Butt: *Minerals of Colorado*, Fulcrum Pub., Golden, 1997.

Embrey, Peter G. and Fuller John P.: *A Manual of New Mineral Names 1892-1978*, British Museum, 1983.

Estavillo, William: *Gems and Minerals of California: A Guide to Localities*, Renaissance House, Frederick, 1992.

Fleischer, Michael and Mandarino Joseph A.: *Glossary of Mineral Species 1995*, The Mineralogical Record, 1995.

Gaines, R.W., Skinner H.C.W., Foord E.E., Mason B. and Rosenzweig A.: *Dana's New Mineralogy*, John Wiley, New York, 1995.

Gramaccioli, C.M.: *Minerali alpini e prealpini*, Atlas, 1975.

Grim Ralph E.: *Clay Mineralogy*, McGraw-Hill, London, 1968.

Hausel, W. Dan and Wayne M. Sutherland: *Gemstones and Other Unique Minerals and Rocks of Wyoming: A Field Guide for Collectors*, Wyoming State Geological Survey, Laramie, 2000.

Haüy, René Just: *Traité de Minéralogie*, Bachelier, 1822

Henry, Darold John: *California Gem Trails: A Field Guide for the Collector of California's Gems and Minerals*, L.R. Gordon, Long Beach, 1957.

Hey, Max H.: *A Second Appendix to the Second Edition of An Index of Mineral Species and Varieties Arranged Chemically*, Trustees of the British Museum, 1974.

———— : *An Index of Mineral Species Arranged Chemically*, British Museum, 1975.

———— : *An Index of Mineral Species and Varieties arranged chemically*, Trustees of the British Museum, 1963.

Kipfer Alex: *Mineral Index*, Ott Verlag, 1974.

Lacroix A.: *Minéralogie de la France et de ses anciens Territoires d'Outre-mer*, Blanchard, 1964.

Ladoo, Raymond B.: *Nonmetallic Minerals*, McGraw-Hill, New York, 1951.

MacFall, Russell P.: *Minerals and Gems: A Color Treasury for Collectors and Guide to Hunting Locations*, Crowell, New York, 1975.

MacLachlan, Don: *Complete Book of Rocks, Gems and Minerals*, Petersen Pub. Co., Los Angeles, 1968.

Mandarino Joseph A.: *New Minerals 1990-1994*, The Mineralogical Record Inc., 1997.

———— : *Fleischer's Glossary of Mineral Species 1999*, The Mineralogical Record, 1999.

Merrill, George P. and Foshag William F.: *Minerals from Earth and Sky: The Story of Meteorites, Gems and Gem*, Smithsonian Institution Series, New York, 1934.

Milovsky, A. V. and Kononov O. V.: *Mineralogy*, Mir, 1985.

National Museum of Natural History: *The National Gem Collection*, Smithsonian Institution, Washington, 1997.

Nickel, Ernest H. and Nichols Monte C.: *Mineral Reference Manual*, Van Nostrand Reinhold, 1991.

Oliver E.: *The Care and Conservation of Geological Material: Minerals, Rocks, Meteorites and Lunar Finds*, Oxford University Press, Boston, 1992.

Palache, Berman and Frondel: *Dana's System of Mineralogy*, Wiley & Sons, 1962.

Pearl, Richard Maxwell: *Cleaning and Preserving Minerals*, Earth Science Pub. Co., Colorado Springs, 1982.

Pellant, Chris: *Rocks and Minerals*, Dorling Kindersley, New York, 1992.

Phillips, William R. and Griffen Dana T.: *Optical Mineralogy: The Nonopaque Minerals*, Freeman, 1981.

Picot P. and Johan Z.: *Atlas of Ore Minerals*, BRGM, 1982.

Ramdohr, Paul and Strunz Hugo: *Klockmanns Lehrbuch der Mineralogie*, Enke, 1978.

——————— : *The Ore Minerals and their Intergrowths*, Pergamon, 1969.

Ransom, Jay Ellis: *Gems and Minerals of America: A Guide to Rock Collecting*, Harper & Row, New York, 1974.

Reynolds, Robert E.: *Minerals from the Mojave*, San Bernardino County Museum Association, Redlands, 2000.

Robbins, Manuel: *Fluorescence: Gems and Minerals under Ultraviolet Light*, Geoscience Press, Phoenix, 1994.

——————— : *Rocks and Minerals*, DK Pub., New York, 1998.

Roberts, Willard L., Campbell Thomas J., Rapp George R. and Wilson Wendell E.: *Encyclopedia of Minerals*, Van Nostrand Reinhold, 1990.

Roberts, Willard L., Rapp George R. and Weber Julius: *Encyclopedia of Minerals*, Van Nostrand Reinhold, 1974.

Rössler Hans J.: *Lehrbuch der Mineralogie,* VEB, 1987.

Sinkankas, John: *Gemstone and Mineral Data Book,* Geoscience Press, Prescott, 1981.

Sofianides, Anna S.: *Gems and Crystals from the American Museum of Natural History,* Simon and Schuster, New York, 1990.

Stalder, H.A., Wagner A., Graeser St. and Stucker P.: *Mineralienlexikon der Schweiz,* Wepf, 1998.

Strunz, Hugo and Nickel Ernest H.: *Strunz Mineralogical Tables,* Schweizerbart'sche, 2001.

Strunz, Hugo: *Mineralogische Tabellen,* Geest & Portig, 1982.

Verbeeck: *Biominerals,* CRC Press, Boca Raton, 1990.

Vozza, Vincent J. and Judith L. Vozza: *Zeolites!: The Misunderstood Minerals,* F. Silva Printing, New Jersey, 1975.

Weiss, Stefan: *Das Grosse Lapis Mineralienverzeichnis,* Weise Verlag, 1994.

William, E.: *Familiar Rocks and Minerals,* Knopf, New York, 1988.

Zim, Herbert Spencer: *Rocks and Minerals: A Guide to Familiar Minerals, Gems, Ores and Rocks,* Golden Press, New York, 1957.

Zoltai, Tibor and Stout James A., *Mineralogy, Concepts and Principles,* MacMillan Publishing Company, 1984.

Index

D

E

F

G

H

I

J

K

L

M

U

V

W

X

Z

□□□